Dedicated to the Memory of

Robert Lincoln Potter

(April 22, 1922 - April 04, 2018)

THE HIDDEN UNIVERSE

What The Cosmos Has Been Telling Us All Along

BRUCE D JIMERSON, BS, MS, JD

THE HIDDEN UNIVERSE

When one tugs at a single thing in nature,
he finds it attached to the rest of the world

John Muir

But when **one accelerates a single mass in nature**
he finds it connected to the rest of cosmos.

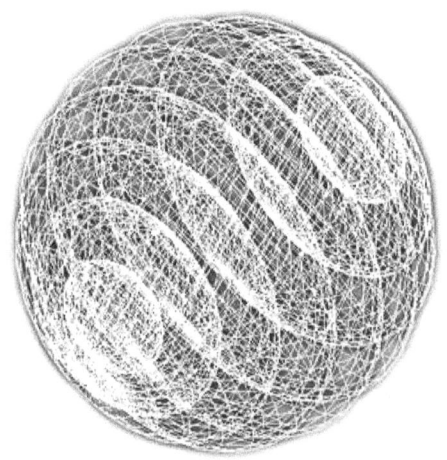

In the grand edifice structured by spatio-temporal evolution, null energy loci share a common volume. From nothing comes exponential exfoliation, thence, the inertial ether upon which gravitational reactance is predicated.

All is Space. The essence of empty space is expansion. The essence of non-expanding space is matter. All forms of matter are spatial contortions. The readiness with which the universe responds to an accelerated mass shows us the hand by which matter grips space and space grips matter. This law of inertia, said Einstein, marks the first great advance in physics.

CONTENTS

Introduction To MG Interdependence..8

Virtual Inertia - The Reality of Missing Mass...13

The Geometry of G Fields 16

Spatial Expansion as a Gravitational Pseudo Force 19

Mach's Principle Reinstated 21
The Effective Hubble Scale 23
The virtual State of Real Space 24
 Energy of Mass 27

Inertia Impedance as Propagation Medium...30
Propagation velocity of Forces 31
Virtual Impedance From Negative Pressure 34

Appendices & Essays

A: Hubble Sphere to Infinite Planes Transformations .....,,,,,. .... .36
B: Circulatory Space Resolves Dark Matter Riddle 37
C: Exponential Expansion Resolves Inflation-Creation Paradigm 38
D: Null Universe Resolves Dark Energy Mandate 39
E: The Inertia of Gravity 40
F: Spatial Inertia Explains gravity 41
G: Spatial Inertia Explains g fields as pseudo Forces 42
H: Convergence Model of **G** as a pseudo Force 43

J: The physical Implication of Einstein's Λ 47
K: Electron Angular Momentum 49
P: Particles as Spatial Circulations 52
Q: Matter From Circulatory Space 54
R: The Fine Structure Constant Alpha..55
S: Spatial InFlux Model of Particle Formation 56
T: SpaceTime Curvature Essay 59
U: Speculations On Unification 60
V: Musings On The Nuclear Force As Linked Spatial Circulations 61
Z. Inertial Reaction as Instant Communication 64

An Introduction to Gravity-Inertial Interdependence

In the grand devise of self creating cosmology, nothing is given nor taken. All substance is illusory, the contortions of micro vorticities created, sustained and acted upon by expansion pressure, even the vortices themselves, are but the machinations of virtual circulations. All is dynamic space and interaction. The reality of virtual Inertia begins with Newton's 2nd Law of Motion.

If the universe is powered by expansion, then the universe must power expansion. Expansion involves momentum flow from space to particles. Particles are therefore negative pressure sinks (irrespective of their polarity). As the building blocks that create the illusion of substantive matter, particles serve as gateway for the return of expansion created momentum. Each particle is a negative pressure sink, and when bound together in large numbers by electrical and quantum forces to form ponderable mass, they exhibit collective inertial opposition to all forms of acceleration.

Fig I-1: All field particles have circulatory spatial extensions. In fact, the particles themselves are constructs thereof. Consequently they bond to one another by linking circulations. This accounts for the elasticity of the bond and the strength of the retention force dependence upon distance, up to a point where the bond is broken. Circulations are not closed rings as indicated, in **Fig I-1**, but rather, every spatial influx momentum stream-tube is an open source of momentum influx. For electrons, each is associated with a mass m_o and an angular momentum $h/4\pi$ as depicted in **Fig I-2**.

Fig I-2 depicts a cross section taken through an imaginary electron in our imaginary unified universe illustrating the physiology underlying the rotational aspect of the angular momentum field $h/4(pi)$ as well as the source of electron mass m_o and charge q_e. Electrons, like all particles having measurable mass, are negative pressure sinks. Momentum Flux flows inward from space. The energy integral taken over the Hubble volume comports with m_o whereas the circulatory spatial field energy corresponds to the angular momentum $h/4\pi$.

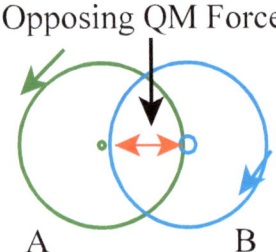

Opposing QM Force

A   B

Electrically neutral particles comprise superimposed clockwise and counter clockwise spatial spin circulations having slightly displaced circulatory centers as shown in **Fig I-3**. Angular momentum spin fields can exists as global electric fields **Fig I-2** or as quantum binding vortices **Fig I-1**, or both in the same particle as is the case with the $\pi+$ pion and π^- pion (wherein net angular momentum is zero because one of two opposite angular momentums is compacted in the binding action of the particle as shown in **Fig I-1**.

Fig I-4 illustrates the neutral particle condition where the electric field of two slightly displaced **h/4π** angular momentum(s) is each nullified by the other, leaving nonetheless a standing energy wave created by the rotation of the vortical centers about a common point midway between, The condition illustrated is that of a neutral particle amalgamated from two equal, but counter rotating, primary particles. The combined mass **M** will be at least twice that of one primary particle. If in **Fig I-3**, **B** represents a positron and **A** an electron, they will combine to create two gamma ray photons unless spaced apart by an interposed quantum condition. .

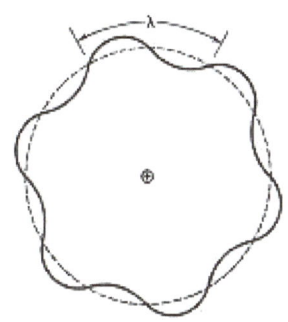

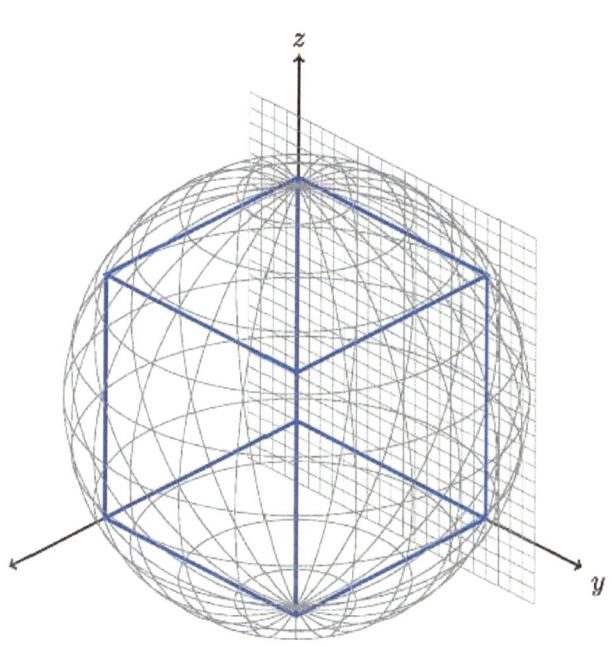

Fig I-5 shows mass **M** at rest in free space at the intersection of the xyz axes. Until a mass decays naturally or slammed apart, it reveals no vestigial evidence of an alter existence as two or more charged particles. Moreover, unless it is within the Hubble universe, it displays no properties whatsoever (no dimensionality, gravity or mass). The decay products of masses include combinations of muons, electrons, positrons, neutrinos and photons. While charge, angular momentum and energy are accounted for by conservation laws, mass is not a conserved quantity. Inertia does not depend upon the particle itself, but rather the universe as a whole. When cosmic mass energy is divided by Hubble area we have, in fact, redrafted Mach's Principle.

Fig I-6 takes a cross section of the cube through the center showing the two halves of the Hubble exerting opposite acceleration forces upon **M** parallel to the **Z** axis. The essence of inertia is discovered in the realization that masses distributed over large volumes cannot act as individual inertial elements or discrete gravity sources, but only collectively as idealized infinite planes having finite area density. As such, the gravity fields thereof are perpendicular to the surface of the plane, ergo the two hemispheres of the Hubble created by slicing through any great circle thereof, act as oppositely directed parallel area density fields. Ergo, net **g** force created by a non accelerating inertial mass **M** = 0.

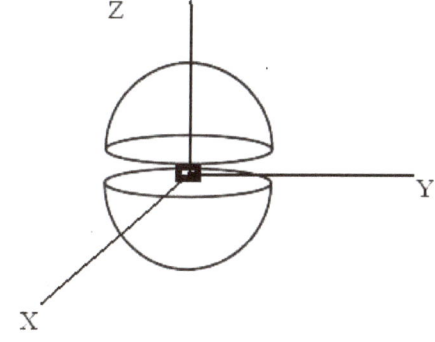

Fig I-7 depicts the **g** field of the Hubble universe as the composite strength of many individual planes perpendicular to the **Z** axis. Each plane is an area density equal to the mass contained in the volume between the planes divided by the area of the plane. Since the **g** field of large planes is perpendicular to the plane, the total **g** field of all planes perpendicular to a particular direction **Z** will be proportional to the Hubble mass M_U divided by the average area of all planes. Herein, the average area density σ_U of the Hubble universe structured from a plenum of planes will be approximated as one kg/per square meter.

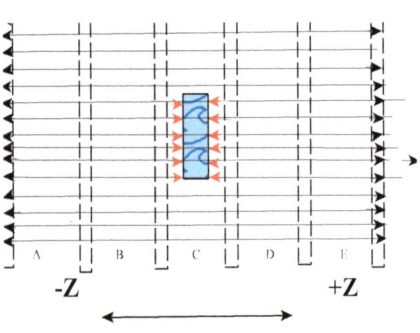

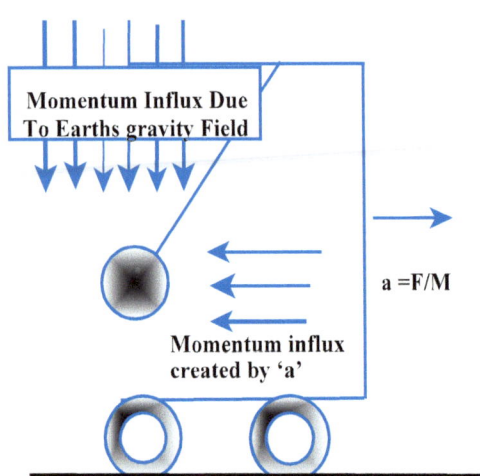

Fig I-8: No experiment can distinguish between force created by gravitational acceleration and that created by inertial reactions, commonly called pseudo forces. One is as real as the other. Both are derived from second law symmetry, the relativism of *"acceleration of mass wrt to the universe"* and *"acceleration of the universe wrt mass."* While the latter would seem to be beyond experiential confirmation, nothing is further from the truth – the isotropic acceleration of the Hubble universe is relentlessly manifest in the gravity field of every mass. From the perspective of a suspended weight on an accelerating platform, the **Ma** flux is no different than the intercepted **Mg** flux.

From Gauss's gravity law, the influx created by a mass M_B in reaction to the exponentially expanding spatial field is $4\pi G \times M_B$. For a uniform spherical mass of radius **r**, the intensity of the **g** field is obtained by dividing by the area $4\pi r^2$, whence:

$$g = \frac{4\pi G M_B}{4\pi r^2} = \frac{G M_B}{r^2} \quad \text{where } G \text{ equals} \frac{c^2}{4\pi R \sigma_U} \qquad (I\text{-}1)$$

For a non-accelerating mass M_B of area A_B, net force is zero in all directions. Nonetheless, M_B/A_B presents a local $(\sigma_B)4\pi G$ spatial expansion deficit wherein pressure will be negative and equal to influx acceleration **g** multiplied by the spatial area density factor σ_U. Whence the pressure **P**:

$$P = g\sigma_U = \sigma_B \frac{c^2}{R} \qquad (I\text{-}2)$$

Pressure is momentum flow, conversely influx momentum $g\sigma_U$ connotes negative pressure, the sink created by the mass M_B. This is the negative pressure at the surface of a uniform density sphere of radius **r** that corresponds to the momentum flow into M_B

Adverting again to **Fig 7**, in lieu of bidirectional spatial expansion parallel to the **Z** axis, the mass M_B is considered accelerating along the **Z** axis. To normalize the area-density of M_B in terms of a common denominator adopted for universe, [σ_U = **one kg/meter²**], the pressure expression (2) can be written as:

$$P = \frac{M_u}{A_U} \times a_1 = \frac{M_B}{A_B} \times a_2 = \sigma_U a_1 = \sigma_B a_2 \qquad (I\text{-}3)$$

Where σ_U expresses the Hubble mass per square meter and σ_B express M_B as mass per square meter. Wherefor, the reactionary acceleration a_1 resulting from an acceleration a_2 is:

$$a_1 = \frac{\sigma_B}{\sigma_U} \times a_2 \qquad (I\text{-}4)$$

That (4) allows a_1 to be greater than a_2 when σ_B exceeds Hubble area density σ_U, poses the question as to how passive Hubble mass creates counter accelerations a_1 exceeding the predicate acceleration a_2. That a_1 can numerically exceed a_2, counter force a_1 cannot be rationally expressed in units of acceleration. When $\sigma_B > \sigma_U$, greater reactionary force is required for a given area. The universe can only exert retro-directive counter force parallel to the **Z** axis. Consequently, the area **A** over which counter force is spread cannot be greater than the projection of M_B upon universe. Counter action a_1 will thus be understood and expressed in terms of it's canonical alter-dimensionality as **ntn/kg**.

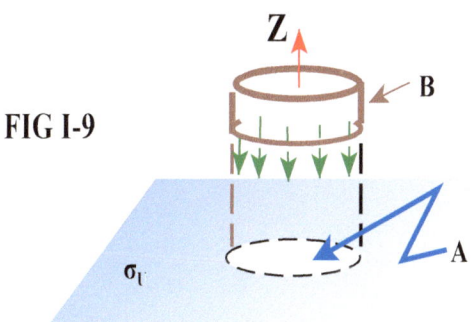

FIG I-9

Fig 9: The action of cosmic inertia impedance σ_U upon the accelerated cylindrical mass M_B creates a reactionary pseudo force field (green) over the projected area **A** of the cylinder upon the universe. That the reactionary force lines are parallel, the intensity of the pseudo force is constant and independent of distance. If the **g** force created by expansion be cumulative and equal to the sum of all area densities, inertial reactance will likewise be cumulative.

While the **g** fields of spheres and point masses fall off inverse square with distance, infinite plane **g** fields are perpendicular to the plane from which they issue. Hence, all **g** force lines from parallel infinite planes are additive and superposed one upon the other. Likewise, every accelerated mass (blue cube) will view the universe as the inertia required to the create the **g** field to the cumulative area density of all lamina into which the Hubble universe might be parsed. In **Fig 10**, green arrows represent the **g** field of expanding space and red arrows are momentum flow into **B** that result when **B** is accelerated to the right as indicated by the blue arrow.

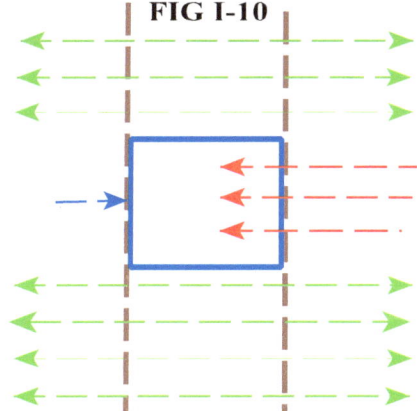

FIG I-10

The essence of Newton's 2nd law, the inertial counter field created by accelerations, can alas be understood in terms of cosmological impedance. All of which leads to a new cosmic coherence, and a restatement of Mach's principle. Henceforth, there is only one law for inertia and gravity, and it serves both for the calculation of inertia from expansion and the calculation of gravity from inertia. Specifically, from (4), virtual cosmic area density σ_U equals Hubble mass M_U spread over an area commensurate with the Hubble manifold $4\pi R^2$ Whence, cosmological pressure is negative and everywhere equal to:

$$-P = \frac{c^2}{R} \times \frac{M_u}{A_u} = \frac{M_u c^2}{4\pi R^3} = \frac{M_u c^3}{3V} \tag{I-5}$$

The revelation to be appreciated in the derivation of (5), is that the c^2 conversion factor betwixt mass and energy is due to the relentless action of spatial expansion upon non expanding matter. The energy of mass is maintained, not by internal forces within, but by spatial expansion from afar. Intrinsic energy, masses have not. Gravitational energy, however, even of the most insignificant mass, is coextensive with the universe -- and that is where particle energy resides. Expansion creates the 'g' field, and the 'g' field defines the energy. The essence of all matter reduces to gravitational field energy created by expanding negative pressure space. Since the energy **E** of a negative pressure volume **V** equals **-3PV**, then total Hubble energy is indeed **--3PV**

$$-3PV = M_U c^2 \tag{I-6}$$

Cosmic energy $M_U c^2$ is created by expansion. The energy of the universe is contained in the spatial expansion pressure (5) that arises for each mass subjected thereto. Hubble acceleration $[c^2/R]$ multiplied by the spatial inertial factor σ_U.

In his 1929 Tribute to Newton, Einstein had this to say:

> *"Every attempt to deny the physical reality of space, collapses in the face of the law of inertia. For if acceleration is to be taken as real, then space must also be real within which bodies are conceived as accelerated."*

Einstein's adaption of empty space as cosmological impedance would, at first glance, appear as an outright *switch-a-roo*. But Special Relativity is theory of mensuration. To the extent it has been verified, an inertial ether is superfluous. Although Einstein never developed the ether paradigm as a numerical quantity, others have. In his later writings he expressed increasing confidence in the affirmation of space as the virtual corporeality upon which inertia and gravity depend. That ethereal impedance is only observed in those circumstances where space and mass are deduced to exist in a state of relative acceleration, is at once a mystery, and a key to unlocking the mystery.

Virtual Inertia - The Reality of Missing Mass

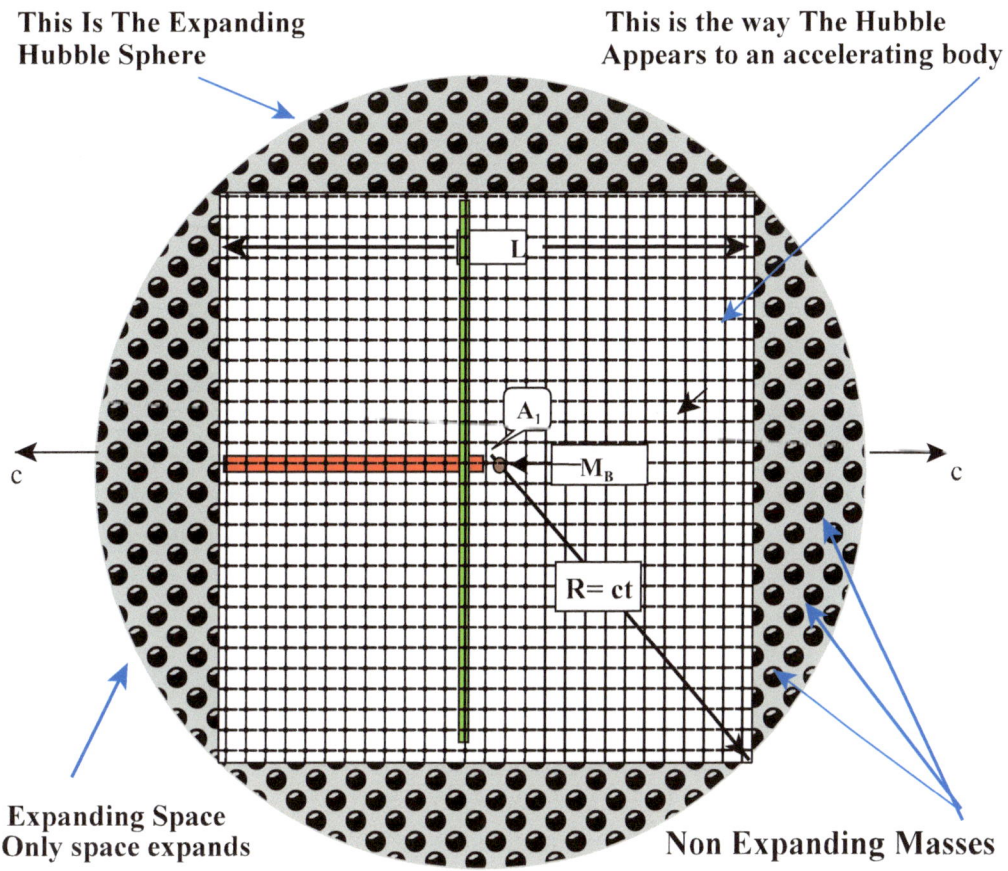

Each spatial volume element of scale 'r_s' expands at an accelerating rate $a_s = H^2 r_s$. The recessional velocity of space during the Hubble time 'τ' at the limit R is then:

$$v = \int_\tau a\, dt = at = \frac{c^2}{R}\left[\frac{1}{H}\right] = \frac{c^2}{R}\left[\frac{R}{c}\right] = c \qquad (1)$$

If the present scale of the Hubble sphere is R, then recessional rate of an object at distance 'r' is:

$$v = \frac{c}{R} \times \frac{r}{1} = \frac{c \times r}{R} = Hr \qquad (2)$$

Commonly misnamed Hubble's Law, but actually first suggested by Howard Robertson, spatial expansion determines the virtual inertial characteristics of the void as well as the ethereal parameters of space as a propagation medium.

Within the known extent of space and mass, the universe is a superlative vacuum. Yet an accelerated or decelerated mass in empty space feels the cosmos as inertial impedance. By what manner can space bring about gravity, inertial reaction and propagation of EM waves?

Surprisingly, the mathematical artifice has long been known, but chronically viewed as of no practical applicability – indeed, the miraculous properties of infinite laminae(s) were considered largely academic. That all change in the latter years of the 20th century. Coming with the 1998 supernova studies, were the imperatives of exponential expansion, potentially infinite space and the problem of finding sufficient energy to fund accelerated expansion.

Cosmological expansion applies to space; the size of material bodies are unaltered, but the mass of material body is not unaffected. The difference in the expansion rate of an empty volume and a volume containing mass is proportional to the mass contained therein. From the perspective of the expanding Hubble sphere, masses are spatial sinkholes. While there is no spatial motion to be observed, the difference between expansion of space and matter (or other forms of mass energy), creates pressure gradients (momentum influx). The misunderstood 'g' field of material objects is simply the manifestation of the negative pressure gradient created by non-expansive matter.

The conflation of space with mass foliates as area-density. While neither Hubble mass nor size is separately utile, taken together, they define the inertial modulus of the cosmos. As the intrinsic impedance of space, the area density operative σ_U is isotropic and ubiquitous. To understand the apparent attraction between masses, one must probe anew the essence of inertia, for that is the facility upon which gravity depends.

Fig 1 depicts the universe as a cubical stack. A one meter cube of density **10^{-26} kg/m^3** will exert a force upon a spherical body **M_B** displaced a distance 'd' from the cube center approx **$G(M)(10^{-26})/d^2$**. Adding more cubes (**B_1, C_1, D_1 ...N**) horizontally (red row) increases the 'g' force, but because 'g' fields fall off inverse square with distance, little is gained by adding additional cubes (no matter how many cubes are added, total force **F_T** is less than twice that of the first cube **A_1** alone, i.e., **[$F_T < [2G \times 10^{-26}M_B/d^2$]**.[1] Suppose instead, cubes are added vertically rather than horizontally (green column). Each new cube adds a small force at a less favorable angle, and because it is also further away from **M**, the effort appears to be even less productive than adding cubes horizontally -- that is, until the number of cubes becomes large –then the lines of force associated with the vertical column are perpendicular to the column. In this construct, an alternative form of Newtonian gravity is provided by Gauss. The acceleration 'g' due to a mass **M** (not necessarily a point mass) can be expressed as:

$$\mathbf{v} = \int_S g \cdot n \, dA = -4\pi GM \tag{3}$$

---

[1]Known as the Basel Problem, originally proposed by Mengoli in 1644 and solved by Euler in 1734:

$$\sum_{n=1}^{n=\infty} \frac{1}{n^2} = \frac{1}{1^2} + \frac{1}{2^2} + \ldots \frac{1}{n^2} = \frac{\pi^2}{6}$$

Fig 2a shows a cube enclosed by a Gaussian sphere of radius 'r.' Average 'g' field intensity diminishes as $1/(4\pi r^2)$. **Fig 2b** (red arrow), depicts a row of cubes enclosed by a cylindrical Gaussian surface **S** of length **L** and radius 'r.' Everywhere along the curved surface of cylinder **S**, the 'g' field points radially inward (anti-parallel to the outward normal unit vector '**n**.' At the flat ends, the inward flux is zero, ergo:

$$g(2\pi rL) = 4\pi GM \tag{4}$$

M is the total mass enclosed by **S** (a segment of length **L** and linear density λ), so it has mass λ**L**, hence, unlike the cube, **g** ∝ **(1/r)**

$$\mathbf{g = (2G\lambda)/r} \tag{5}$$

Considering next cubes added to each side of the green column to form a plane **(Fig 2c - red arrow)**. If the area density of the plane is **σ** (kg/meter²), then the imaginary Gaussian surface will be pill box in shape, where the flat faces have an area '**A**' parallel to the surface of the plane as shown. The '**g**' field is everywhere perpendicular to the end areas, so the curved sides of **S** contribute nothing to the integral, hence

$$\int g \cdot n \, dA = -4\pi GM \tag{6}$$

The integral in this case is just the area of the two ends of the pillbox cylinder, 2A, hence:

$$\mathbf{-g_S(2A) = -4\pi GM} \tag{7}$$

Mass **M** is simply the total amount enclosed by the surface **S**, which is analogous to a cookie cutter punch-out of a circular area of the plane having mass $\sigma_S A$, hence:

$$\mathbf{-g_S(2A) = -4\pi G(\sigma_S A)} \tag{8}$$

Therefore:

$$\mathbf{g = 2\pi GA} \tag{9}$$

The '**g**' field of a large uniform density plane is perpendicular to the plane, ergo, field lines are parallel and consequently, field intensity is constant with distance. For an infinite plane, the ends of the pillbox can be located at any distance from the plane without affecting the result. Two parallel infinite planes having the same area density σ_S would produce twice the gravitational intensity as one plane irrespective of the separation distance between the planes. As more parallel planes are added, the '**g**' field increases proportionately. For the entire cube (**Fig 2d**), the local gravitational condition for an internal mass $\mathbf{M_B}$ will always be net zero as it resides at the center of its own Hubble sphere.

If the cube represents the universe (**Fig 1**), it is known by experiment an accelerated mass within the universe experiences a counter reaction. It is also known that the '**g**' field will be net zero in every direction. The addition of parallel laminates to build the **Fig 2d** cube from **Fig 2c**, enhances the external '**g**' field, but the internal '**g**' always cancels at the center of the cube (**Fig 4**).

G Field Geometrics

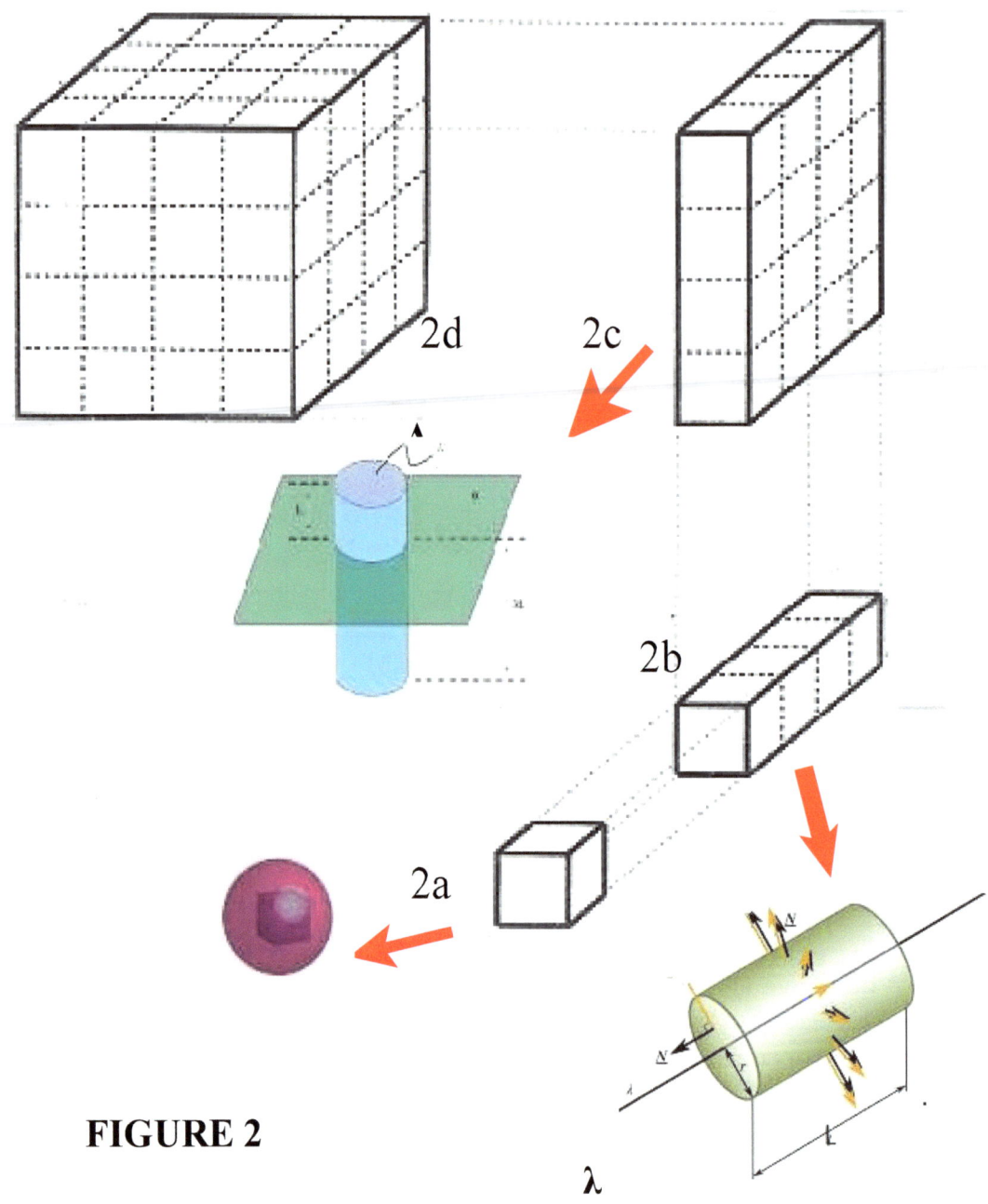

FIGURE 2

Unidirectional '**g**' fields induced within a stationary mass in an accelerating universe, are no different than the reactionary forces apprehended by accelerating bodies in a non accelerating universe. Gravity is cosmological testament to Einstein's "*Principle of Relative acceleration.*"

The cumulative inertia of all planes comprising the cube is the virtual inertia of the cube that exist at all locations within the cube. Virtual inertia, like the **g** field of the ensemble, is ubiquitous. That the strength of the cancelling gravity fields at the center depends upon the number of laminates on each side of center, they act as a tug of war on an in-between mass. Gravity fields are reactionary pseudo forces, strength depends upon the mass marshaled to bring about the reactance. Consequently, the inertial reaction at any location within the cube depends upon every laminate participating in creating the cancelling '**g**' fields (**g** field pressure of planes to the left are cancelled by **g** field pressure of planes to the right), but the inertia that created the field does not cancel. The left side field is equal to the total area density of all planes to the left multiplied by the expansion acceleration in the (-**x**) direction and the '**g**' field to the right is equal to the total area density of all planes to the right multiplied by the expansion acceleration in the (+**x**) direction. Any acceleration of M_B creates an imbalance - there is no difference between the force created by the acceleration of the universe and that created by the acceleration of M_B. Per **Fig 2**, point mass intensity falls off as $1/r^2$, for a line of mass having density λ, intensity falls off as $1/r$, and for a plane of area density σ_P, gravitational intensity is constant and equal to σ_P at any distance. A cube constructed of '**r**' laminae, each of area density σ_P, will manifest virtual density σ_U at all locations equal to **r**σ_p

Gravity like inertial reaction, is proportional to the inertial mass M_B of the object from whence it derives. For an accelerated mass, reactiony force lines are opposite to the direction of acceleration. A force created by unidirectional acceleration of the universe with respect to a mass M_B at rest is indistinguishable from a force created by unidirectional acceleration of M_B wrt the universe.[2] The apparent emanation of gravitational flux from an inertial mass M_B is the pseudo force manifestion of the distortion imposed by M_B upon the isotropic spatial expansion field. Physiologically, **g** fields are momentum influx gradients created by the difference between the expansion within material masses and the expansion rate of empty space. .

As there is an Archimedean buoyancy relationship between the integral of pressure over a surface and *weight-of-fluid-displaced,* an analogous force is afforded by Newton's 2nd law. Both Archimedean and Newtonian reactions are unidirectional. Newton's 2nd law takes into account total mass just as Archimedes's Principle outputs buoyancy force as displaced fluid weight. For spherical bodies having approximately uniform density (most planets, moons and stars), '**g**' fields can be related to spatial expansion surface pressure. From a Theorem due to Gauss, volumetric density ρ of '**B**' transforms to area-density σ_B, and, volumetric cosmic density transforms to area-density σ_U. For exponentially expanding space, cosmic divergence $a_n = c^2/R$. The pressure (P_s) on the shell area-density of **B** is:

$$-P_B = a_n \sigma_B = \frac{c^2}{R} x \frac{M_B}{A} = (-g\sigma_U) \qquad (10)$$

---

[2] Einstein's doctrine of relative acceleration. Asimov, Issac, *Understanding Physics,* Barns & Noble, 1966 Chapter VI at page 118.

Per (10), gravity fields can be specified either as pressure **P** or acceleration **g** (wherein $\mathbf{P_B}$ denotes momentum influx pressure (**ntn/m²**) on the surface of a uniform spherical shell of mass $\mathbf{M_B}$). To express gravity in the customary units of acceleration (**ntn/kg**), both sides of (10) are divided by σ_U. Acceleration '**g**' at the surface of **B** is then expressed in terms of ($\mathbf{P_B}$) and $\mathbf{a_n}$. [a unity value σ_U equal **1 kg/m²** contributes to dimensionality but not magnitude]. That σ_U = **1 kg/m²** specifies the inertial property of the universe as a necessary factor in reconnoitering traditional dimensionality of gravity as acceleration, it will also be understood as vital to the conceptualization of gravity and inertial reaction as pressure fields. Newton's formulation of gravity as acceleration offers no hint as to its cause. Pressure is identified with momentum flow, that which arises from the difference between the expansion rate of free space and non-expanding matter. What is sensed as gravity on the surface of $\mathbf{M_B}$ is the apex of the negative pressure gradient (**Fig 3**).

To effectuate as a force, spatial influx must carry momentum. In transforming from surface pressure (**ntn/m²**) to acceleration (**ntn/kg**), local gravitational fields are intrinsically imbued with the energy density function σ_U. By expressing the ratio of the normalized area densities (σ_B of **B** and σ_U of **U**), the sublimity of cosmic mass as a participant in the formulation of local '**g**' fields (while not obvious from Newton's expression for gravity) is revealed in the Gaussian formalism.

Gauss's Law of Gravity impregnates σ_U as an adjunct of accelerating space. Understanding Newton's 2nd law as the inertial response of the universe is key to understanding Newton's "*Law-Of-Gravity.*" Accelerating space is the momentum communicating medium. Gravity fields are the manifest of spatial influx created by matter - every mass is a negative pressure sink.

Gravity fields continuously transport momentum – herein that function is expressed as the pressure term formed by the (**g**)(σ_U) product per (10) above. Specifically, if (**g** x σ_U) is momentum flow entering $\mathbf{M_B}$ and ($\mathbf{a_n}$ x σ_B) is momentum deficit created by the impedance of $\mathbf{M_B}$, then:

$$g = \frac{\sigma_B}{\sigma_U} a_n \qquad (11)$$

Wherein both σ_U and σ_B are normalized as area-density per square meter. Gravity is thus a space based Newtonian pseudo force as first remonstrated by Richard Feynman.[3]

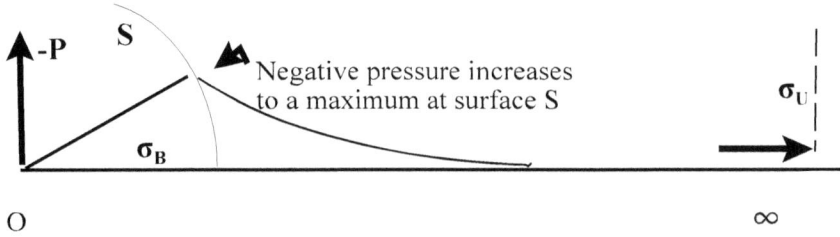

---

[3]"*One very important feature of pseudo forces is that they are always proportional to the masses. The same is true of gravity. The possibility exists therefore, that gravity itself is a pseudo force. Is it not possible that perhaps gravitation is due simply to the fact we do not have the right coordinate system?*" [**Feynman – Lectures on Physics, Vol I at 12 - 11**]

A Physical explanation of gravity from expansion

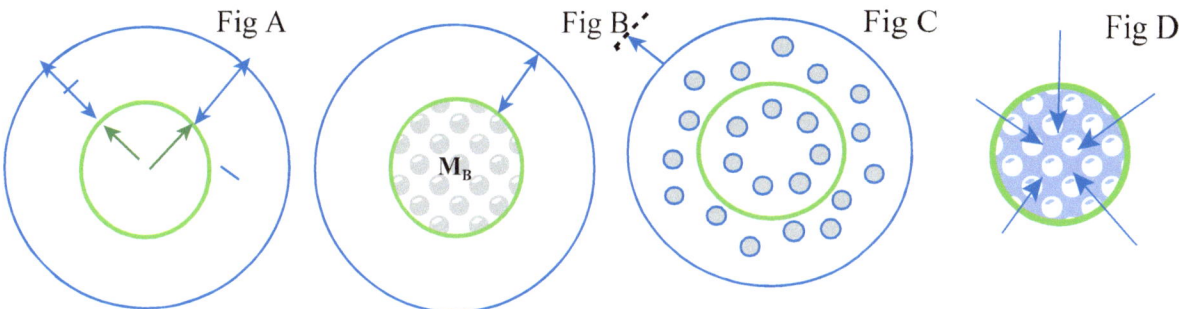

Fig 3A. Green circle defines an accelerating volume of empty space expanding within a larger volume of expanding space (blue).

Fig 3B. The inertial mass of gray spheres populating the green volume impairs spatial expansion. This leads to two possibilities. If the spheres are unbound and sufficiently separated, they will radially diverge with the flow of space from the green volume into the blue volume per Fig 3.

Fig 3C. Expansion of space within green volume causes unbound masses to diverge radially outwardly into empty surrounding blue volume. Dotted black line indicates spatial volume now occupied by the blue volume. Gray spheres represent galaxies.

Fig 3D. Spheres are bound together (simulated by blue background which constitutes a binding force). The weakness of the expansion acceleration produces no observable change. The inertial property of the non expanding assembly impedes internal spatial expansion. Space carrying inertial momentum flows inward (blue arrows). The rate of inflow corresponds to the negative pressure differential between the internal and exterio expansion factors. Blue arrows thus represent momentum flow associated with the pressure differential at the green surface. Diminution of momentum flux with distance delineates the gravitational gradient. Newton called momentum influx "*gravity*," but neither Newton nor Einstein recognized the field as an **F = ma** pseudo force. That distinction would be left to Richard Feynman.[3]

Each mass is considered the agent of its own gravitational pressure deficit. An alternative interpretation views mass as highly condensed space (contracted space). Momentum influx supplies the energy by which every mass sustains its **Mc²** energy status. On the cosmic scale, creation of momentum by expansion equals absorption of momentum by mass. Mass grows in energy as the integral of absorbed momentum?

$$E = \int mv \, dv = \frac{mv^2}{2} \tag{12}$$

Mach's Principle Reinstated

Intuitively, momentum flux created by expansion of free space might be expected to equal momentum flux absorbed by inertial matter. That cosmic mass acts collectively as an inertial area density, the confluence of the Hubble sphere with its parts, is nothing less than a confirmation of Mach's Principle. Total cosmic mass is indeed the determinative factor in the specification of σ_U, the long dormant Principle attributed to the 19th Century physicist, Ernst Mach, is alas validated. The universe opposes accelerated bodies with the same force as that required to create the acceleration. For unidirectional acceleration, the anti-parallel responsive action exerted upon each elemental volume of mass is retro-directively aligned therewith. Only that part of the universe orthogonal to the projection of the mass of the accelerated body upon the inertial plane of the universe, is involved in creating unidirectional counter action -- and then solely to the extent of the projected area.[4]

Fig 4 depicts a slice taken through the **X-Y** plane of the universe **U** depicted as two side-by-side slabs U_1 and U_2 each having area density $\sigma_U/2$. Each slab is considered a plurality of '**n**' parallel planes. A thin uniform density rectangular plate **B** at rest midway between U_1 and U_2 experiences the '**X**' component of the '**g**' field (that created by the outward acceleration of the two slabs) as perpendicular to each face (red arrows). The '**g**' field from U_1 penetrating the left face of M_B equals the '**g**' field from U_2 penetrating the right face of M_B. The '**g**' fields of the U_1 planes equals the '**g**' fields created by U_2 planes. Mass M_B tugs upon the left and right half of the universe equally. Emphasizing again, the '**g**' fields of large flat planes are perpendicular to the plane, the '**g**' field created by all planes in each slab will be normal thereto, consequently force lines are parallel and only that part of each plane lying within the projection of M_B upon the planes (dotted blue lines) is effective in creating a '**g**' force (parallel to the **X** axis) that will act upon the plate **B**.

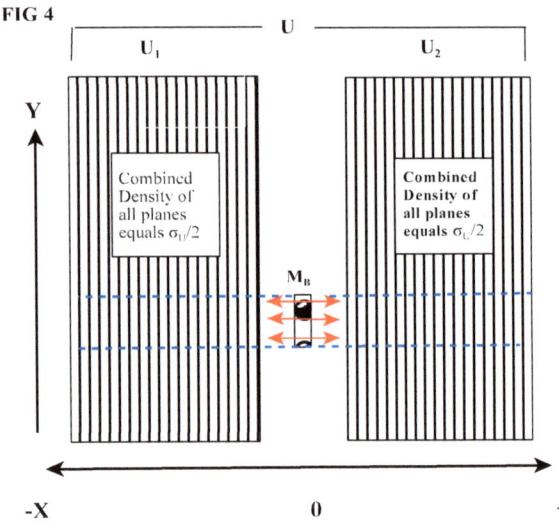

FIG 4

Red arrows indicate pressure created by planes comprising U_1 equals pressure created by all planes comprising U_2. Each half of the **U** universe exerts an expansion force per unit area A: Because '**g**' fields of ∞ planes are ⊥ to the planes, the cumulative expansion force exerted by U_1 and U_2 planes upon an at-rest mass M_B is:

$$F/A = P(M_B)[(\sigma_U)/2][(a_n)-(a_n)] \qquad (13)$$

where a_n is relative acceleration between the universe and a non accelerating mass M_B.

---

[4]The concept of gravity and inertial reaction as continuum pressure created pseudo forces abrogates the problems posed by individual lines of force or the conveyance of momentum in quantified units. There may in fact be no such thing as quantize units of force capable of acting upon individual atoms or parts of atoms. Whether a minimum unit of gravitational force exists is of no consequence from the perspective of a gravitational pressure continuum.

When 'B' is accelerated in either direction along the horizontal 'X' axis, the relative acceleration between M_B and the universe is:

$$P_B = \frac{F_B}{A_B} = \ddot{x}\sigma_U = \frac{M_B(a_B)}{A_B} = \sigma_B(a_B) \tag{13}$$

Consequently:

$$\ddot{X} = a_B \frac{\sigma_B}{\sigma_U} \tag{14}$$

In words, (14) expresses Newton's 2nd Law as the acceleration factor [**X double dot**] that corresponds to the reactionary force per unit mass [**ntn/kg**] arising when masses are accelerated. wrt the universe. The area densities σ_B and σ_U of mass '**B**' and the universe '**U**' respectively, are normalized by the common area of interaction A_B (the projection of M_B on the inertial area density factor σ_U of the universe (dotted blue lines). That σ_U has unity dimensionality in the mks system of units, *Mach's Principle* enters the equation in a manner that appears almost providential. The fact of course, is that this is the only formulation possible within the physical laws of global conservation.

In the light (14), the notion of the universe as an "infinite-plane, retro-directive area-density," resolves the effronteries charged against Mach's Principle *by Albert Einstein.*[5] Indeed, Einstein's prescription for an instantaneous inertial medium is aptly satisfied by the ubiquitous virtual area-density $\sigma_U = 1$ kg/m^2. This in essence, is the inertial ether Einstein returned to in his address at Leiden University. In the light of the instantaneous nature of Newton's second Law, the universe has no option to act otherwise. Having dissected the Hubble into thin planes, we then analyze how the inertial force must accumulate with the addition of each new g creating parallel plane. From this we explain the mass dependent gravity force that is augmented with each new laminate. A body M_B sandwiched between the two sets of laminations (**Fig 4**) that represent the expanding Hubble sphere, feels the cumulative **g** field, which is zero -- unless M_B is itself accelerating. There being no distinction between **g** field accelerations and inertial reaction accelerations, net force upon a mass in motion is always a combination thereof. For an accelerating mass M_B, the universe appears as a finite cumulative area density σ_U commensurate with a finite cosmic depth. . .

---

[5] In has address at Leiden University in 1920, Einstein remonstrated:

"It is true that Mach tried to avoid having to accept as real something which is not observable by endeavoring to substitute in mechanics a mean acceleration with reference to the totality of the mass of the universe in place of an acceleration with reference to absolute space. But inertial resistance opposed to relative acceleration of distant masses, presupposes action at a distance; and as the modern physicist does not believe that he may accept this action at a distance, he comes back once more to the ether, which has to serve as a medium for the effects of inertia. But this concept of the ether to which we are lead by Mach's way of thinking differs essentially from the either conceived by Newton, by Fresnel and by Lorentz. Mach's ether not only conditions the behavior of inert masses, but is also conditioned in its state by them"

The Effective Hubble Scale

For conservation of momentum, an acceleration of **B** wrt the universe must be countered by an acceleration of the universe in the opposite direction. Oppositely directed accelerations of two masses are opposed by the universe. Suppose **B** is split in to two halves and each half is propelled in the opposite direction as shown in **Fig 5**.

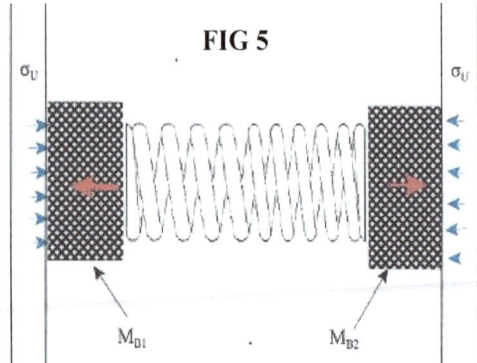

FIG 5

Cosmic area density σ_U acts at all locations to oppose accelerations over an area defined by the projection the accelerated mass upon the σ_U plane $\perp$ to the acceleration. If A_{B1} and A_{B2} are the projections of the masses shown in **Fig 5**, the inertial pseudo forces will be:

$$a_1 \frac{M_{B1}}{A_{B1}(\sigma_U)} = a_2 \frac{M_{B2}}{A_{B2}(\sigma_U)}$$

Since σ_U cancels on both sides, the pressures will be equal when: $\quad a_1 \sigma_{B1} = a_2 \sigma_{B2}$ \quad (16)

Both densities are normalized by common denominator (one square meter), ergo, reactionary acceleration is consistent with what is already known from Newton's Third Law, equal, opposite and along the same line of action. The universe (infinite plane) is retro-directive.

$$a_1 M_{B1} = a_2 M_{B2} \qquad (17)$$

That Newton's second law is derivable as virtual acceleration pressure, simplifies the requirement that each element of an accelerating mass separately communicate with the universe *a la* some a force transmitting particle (e.g. gravitons). Inertial space is the virtualization of D'Alembert's perfect in-viscid fluid. In the void, Newton's laws of motion apply, yet no force is to be found unless the relative velocity between space and mass is changing either in magnitude or direction. Momentum flow (spatial influx) equals pressure differential at the space mass interface.

Negative spatial pressure has no known static properties. It need not obey Pascal's law. While spatial pressure as momentum flow is metaphorical, the behavioral characteristics of space are in accord with those of Bernoulli. Lower negative pressure means higher flow gradient.

Energy in the form of mass carries with it the angular momentum fields of its parts. All such energies are opposed by the characteristic impedance of the inertial acceleration modulus σ_U = **kg/m²**. As an isotropic and ubiquitous property of space, σ_U conceptualizes as the virtual area density created by superposition of a plurality of infinite planes having area-density approximately equal to the Hubble matter content spread over the Hubble sphere. From the perspective of the energy available to create an energy density **kg/m²** taking a Hubble sized sample of the universe, the shell area as operative plane will be smaller than the Hubble sphere by a factor of **5/6** (Appendix A). Taking $H_o \approx 70$, $R_H = 1.3 \times 10^{26}$, then the effective radius for calculating area and density is:

$$R = (5/6)R_H = (1.3 \times 10^{26})(5/6) = \mathbf{1.08 \times 10^{26}} \text{ meters} \qquad (18)$$

The Virtual state of Real Space

As depicted in **Fig 4**, the **g** field of all left side and right side laminates marshal as single slabs U_1 and U_2 respectively. For a mass M_B therein, equal and opposite negative pressure fields arise (red arrows). Assume M_B represents the earth and with its mass flattened into an area equal to the earth's surface. From (13), the acceleration acting upon the surfaces of a non moving mass M_B are those created by spatial expansion parallel to the **X** axis, accordingly:

$$P_B = \sigma_B a_B = \frac{M_B}{A_B} a_n = \frac{5.98 \times 10^{24} \text{kg}}{[6.37 \times 10^6 \text{m}]^2} \times \frac{c^2}{4\pi R} \quad (19)$$

From (18), **R** is taken as 1.08×10^{26}, thence:

$$P_B = [0.1474 \times 10^{12}] \times \frac{9 \times 10^{16}}{12.56 \times 1.08 \times 10^{26}} = 9.76 \text{ ntn/m}^2 \quad (20)$$

From (14), there is thus good reason to believe that σ_U is equals **one kg/meter²**. In which case:

$$\ddot{X} = g = \frac{\sigma_B}{\sigma_U} \times \frac{c^2}{R} = 9.76 \text{ m/sec}^2 \quad (21)$$

The '**g**' field on either surface of **B** can be transformed away by simply accelerating **B** in either the (**- X**) direction or (**+ X**) direction. Eliminating the '**g**' field on one side by accelerating **B** doubles the reactionary field on the opposite side. The doubling of the '**g**' field one side is really due to the acceleration of **B** relative to the universe, a 2nd Law reaction.[6] Just as the cosmic **g** field distends cumulatively across the entire laminae of planes into which the universe could be divided, so also is the inertial area density cumulatively present as σ_U at all locations and in every direction.

Per (11) and (13), every element of mass comprising the matter content of an accelerated body experiences a passive retro-directive inertial counteraction. Space represent the universe in the form of an imaginary fluid having a reactive inertial area density σ_U = **one kg/m²**. Expressed in terms of the area density of the aggregate mass lying within the projection of an accelerating body **B** upon cosmic area density σ_U of the plane normal to the direction of acceleration, then for a flat plate having area density σ_B = **one kg/m²** accelerated normal to its surface at a rate a_1:

$$a_2 \sigma_U - \sigma_B a_1 \quad (22)$$

And since the acceleration of **B** wrt the universe is equal to the acceleration of the universe wrt **B**, $a_2 = a_1$, and therefore $\sigma_U = \sigma_B$ = **one kg/m²**.

---

[6] A small mass will have a small '**g**' field because the cosmic acceleration factor c^2/R is on the order of 10^{-10} meters/sec². Ergo, unless there is a large mass in the vicinity, the '**g**' of a small mass will be totally obscured. So while both gravity and Newtonian reactions are pseudo forces, the former become important only where large masses are involved.

The condition σ_U = one **kg/m²** is imposed upon the projected area of the accelerated body by the isotropic inertial capacity of the cosmological fluid. By this means, the common area counter pressure (that defined by the projection of the accelerating mass upon the plenum of planes that comprise the inertial impedance of the universe along the line of action of the acceleration) is transformed from *force-per-unit-area* to its 2nd law form as *force-per-unit-mass*.

Fig 6 illustrates the σ_U function as a passive retro-directive area density existent at all locations, but undetectable accept at those places where elemental units of mass reside - and then only when the body as a whole is accelerated. As depicted, the effective pressure plane acting upon an accelerating 3-d body is, in general, not confined to a single 2-d surface. Ubiquitous and ever present at all locations, thence instantly asserting opposition to each and every element of an accelerating mass. To understand local inertial impedance, it is necessary to account for the universe as a whole. Things are the way they are, because they cannot be otherwise.

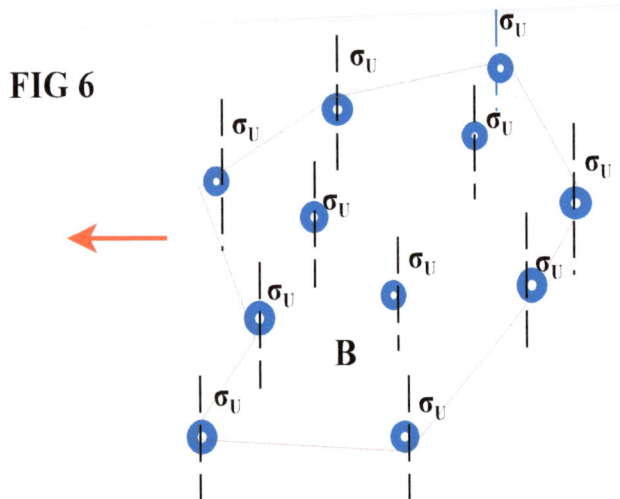

FIG 6

Fig 6: Body **B** (consisting of 11 elemental masses linked by internal quantum forces), is accelerated in the direction of the red arrow. While non accelerating bodies experience the universe as a near perfect vacuum save for the influences of thinly scattered matter throughout the void, all accelerating masses experience space as a superimposed virtual area-density σ_U. Each particle is instantly opposed by the operative isotropic impedance σ_U. While space, from an inertial perspective, can be emulated as hydrodynamic impedance, it differs from normal fluids in that pressure is negative.

Space is *sui generous*. The long sought quintessential spatial medium emerges with the countenance of a fully functional inertial fluid. All trace of its infinite plane pedigree are merged within a ubiquitous isotropic area density modulus σ_U. Although a single functionality, σ_U plays different roles in various guises. For accelerated masses, space is an elastic mechanical conservator of energy. As a medium for transverse wave propagation, it provides supporting impedance upon which transverse wave velocity depends. As the inertial coupling factor between gravitationally attracted masses, spatial flu emulates as momentum flow.

In previous works, σ_U has been approximated by transforming Hubble 3-sphere bare mass ($M_U \propto 1.5 \times 10^{53}$ **kg**) to an equal mass-energy 2-sphere shell model (having an effective radius **1.08 × 10²⁶ meters**). Because 2-sphere gravity energy is less than the 3-sphere gravity energy by a factor 5/6, the 2-sphere radius must be less by the same factor to create the same operative area-density:

$$\sigma_U = M_U/4\pi(R_2)^2 = (1.5 \times 10^{53} \text{ kg})/12.56 \times (1.08 \times 10^{26} \text{ meters})^2 \approx 1 \text{ kg/m}^2 \qquad (23)$$

From a dimensional perspective, σ_U follows from the operative implicate of Newton's 2nd Law, specifically if spatial inertia in the form of Hubble area density M_U/A is the agent of impedance opposing acceleration of individual masses wrt thereto, then for a body **B** of mass M_B,

$$F = M_B(a_1) \tag{24}$$

If **F** represents the passive reactive force of the universe, then if both sides of (24) are divided by **kg**:

$$a_2 = \frac{F}{kg} = \frac{M_B}{kg} \times a_1 \tag{25}$$

In words, (25) states the reactance of the cosmos a_2 to the acceleration of a mass M_B at a rate a_1. This is the acceleration form of Newton's 2nd law, however, a_2 need not refer to a change in velocity. As the reactive agent of the cosmos, a_2 can also be expressed dimensionally as **ntn/kg**. One numerical value with conceptually different mensuration units and deep consequences, in particular the acceleration of **B** at a rate a_1 wrt the universe will not in general equal the acceleration a_2 of the universe wrt **B**. To obtain the pressure form of Newton's 2nd law, both sides of (25) are divided by square meters:

$$\frac{1}{m^2} \times a_2 = \frac{1}{m^2} \times \frac{M_B}{kg} \times a_1 \tag{26}$$

Thence (26) can be written:

$$\frac{kg}{m^2} a_2 = \frac{M_B}{m^2} a_1 \tag{27}$$

The left side of (27) is counter pressure exerted by the universe whenever a mass is accelerated. Being confined to the effective common area defined by the projection of mass M_B thereon, the pressure exerted over the common area of coupling between the universe and an accelerated mass **B** will be equal to the pressure exerted by the universe upon **B**. If M_B is one **kg**, then a_2 will equal a_1. For any other value of M_B, $a_2 \neq a_1$.

The area density factor **kg/m²** is epistemologically consistent with the abductive inference drawn from cosmological parameters. Dimensional derivation *a la* Newton's 2nd law resolves a long standing mystery as to why the ratio of Hubble mass to Hubble area so nearly equals unity in the **SI** system. As previously, (27) can then be written as:

$$a_2 \sigma_U = a_1 \sigma_B \tag{28}$$

As applied to inertial reaction, (28) reveals how the universe functions as a reflective inertial impedance in opposing acceleration of inertial masses. When $a_1 = c^2/R$, then (28) restates the gravity equation where $a_2 = g$ and σ_B is the area density of **B**.

The Energy of Mass

Equation (29) was initially synthesized from General Relatively by Alexander Friedmann without the cosmological constant. The later inclusion of Λ was instrumental in the formulation of the present ΛCDM model. The introduction of (29) in this development, however, is owed to the fact an identical expression can be arrived at obliquely using only the energy relationships applicable to a free falling body in a gravitational field.[7] The once disparaged $\Lambda R/3$ acceleration is now reckoned as spatial exponentiation, the liaison between negative energy space and positive energy matter.[8]

$$\ddot{R} = -\frac{4\pi G}{3}\left[\rho_u + \frac{3P_s}{c^2}\right]R + \frac{\Lambda R}{3} \qquad (29)$$

Self creating cosmology requires net energy be zero on the global scale. This condition is perpetually assured when spatial pressure P_s is negative and equal to:

$$-P_s = \rho_U c^2/3 \qquad (30)$$

In which case (29) simplifies to:

$$\ddot{R} = \Lambda R/3 \qquad (31)$$

Rewriting (30) in terms of Hubble mass M_U

$$-P_s = \frac{M_U(c^2)}{4\pi R^3} = \frac{M_U}{4\pi R^2} \times \frac{c^2}{R} = \sigma_U \times \frac{c^2}{R} \qquad (32)$$

And the energy E in a volume V of spatial pressure P_s is:

$$E = 3P_s V = \frac{3c^2}{R} \times \frac{4\pi R^3}{3}\sigma_U = 4\pi R^2 c^2 \sigma_U = M_U c^2 \qquad (33)$$

Negative pressure energy created [per (32)] by accelerating expansion is manifests as positive energy per (33). The factor c^2 that relates mass-energy to the negative pressure field is due to expansion. Einstein's relationship between energy and mass ($E = Mc^2$) can now be understood for any M to be the consequence of dynamically created spatial expansion pressure. Inertial matter M is but a reflexion of negative g field pressure. Without expansion there is no pressure, without pressure there is neither inertial-mass nor mass-energy. That $4\pi R^2 c^2 \sigma_U = M_U$, see infra (72).

---

[7] As shown by British Cosmologists, William McCrea and Edward Milne (circa 1934).

[8] The 1998 supernova studies revealed the experimentally determined expansion factor concordant with Einstein's cosmological factor $\Lambda R/3$.

Einstein had determined the cosmological constant Λ would need to have a value of $3H^2$ to cancel gravity. From (31):

$$\Lambda R = 3H^2 R = \frac{3\dot{R}^2}{R} \qquad (34)$$

Whence:

$$\frac{\dot{R}^2}{R^2} = \frac{\Lambda}{3}$$

Then:

$$\frac{\dot{R}}{R} = \left[\frac{\Lambda}{3}\right]^{1/2} = H$$

And therefore

$$R = e^{Ht} \qquad (35)$$

A universe with equal positive and negative energy has the same solution as de Sitter's empty universe.[9] Because negative and positive energies are created equally during expansion, no dark energy is required, spatial exponentiation is self creating. The zero energy universe need not have a temporal beginning, existence is past eternal and possibly inevitable. *"Time"* need not exist as a separate dimensionality having defined rate of flow.[10] That change is self perpetuating, it may also be self creating. The inertial mass-energy of individual bodies increases proportionately with the scale factor **R** whereas cosmic energy increase as the square of the scale factor **R**.

For an accelerated mass **M**, the universe responds over a finite area defined by the projection of **M** upon an imaginary plenum of orthogonal area-density laminae orthogonal to the acceleration. That the σ_U factor is finite and isotopic, indicates that cosmic matter content is limited to a volume having a scale comparable to the Hubble sphere. This leads to two or more possible interpretations.

1) Only those objects within the traveled distance of presently arriving photons contribute a cosmological influence to the energy and inertial properties of mass.

2) The nebula are not comoving with exponentially expanding space -- they are wafted apart by the kinematic recessional momentum of accelerating inertial space.[11]

---

[9] First studied by Willem de Sitter, circa 1917.

[10] What is the difference between a *"timeless"* universe and a universe devoid of change?

[11] Appendix B the standard candle interpretation of **1a** super nova

Inertial Impedance As Propagation Medium

The velocity distance law (2) follows from the isotropic homogeneity of space. While the phrase *"empty space"* is meaningless and without physical connotation, it nonetheless conveys a kinematic state describable in terms of virtual parameters (pressure, inertia and acceleration).

If Hubble volume has mass M_U as a 3-d sphere, it will not have mass M_U when the masses are configured into flat lamina (**Fig 4**) or concentric shells (**Fig 7**). Like a plenum of parallel planes, the nested set configuration is cumulatively reinforcing, virtual area density impedance is reduced by an order of one dimension. More specifically, the mass of each shell of radius '**r**' is distributed over an area $4\pi r^2$. Consequently inertial opposition to acceleration is reduced by a factor of 3 and increased by a factor **R**).

$$M_U = \rho_U \int_R \frac{4\pi r^2}{1} \, dr \Rightarrow \sum_1^n \frac{4\pi r^2}{1} \sigma_U \qquad (36)$$

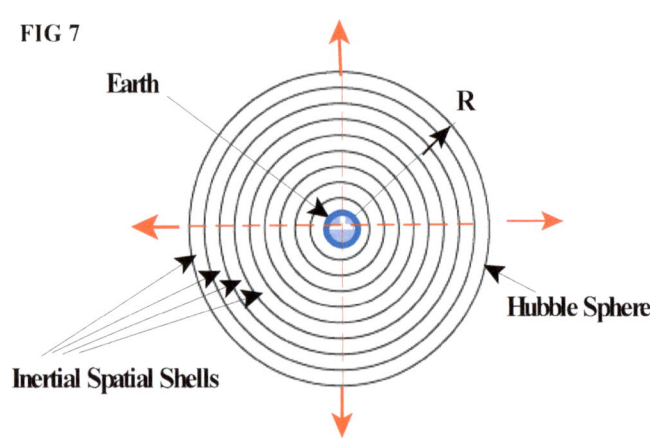

FIG 7

Fig 7: The reaction of earth's inertial mass to isotropic spatial expansion (red) creates a negative pressure sink which absorbs diverging space (**Fig 3-D**) as $[g(\sigma_U)]$. That local inertia can be successfully related to global inertia, the (σ_U) factor must = **1 kg/m²**. In this restated implicate of *Mach's Principle*, any other causal determinative would contradict Newton's 2nd law. Suffice to note, cosmic counter action marshals the cosmic inertial content in the form of pressure $(\sigma_U)g$.

As a composite, the shells act as a single 3-sphere. As individual shells that collectively contribute to the composite **g** field, each shell portends an area density adding cumulatively to the **g** field of every other shell. Consequently area density is summed at all locations. Integration of (36) gives:

$$\rho_U \left[\frac{4\pi R^3}{3} \right] = \sigma_U 4\pi R^2 \qquad (37)$$

Then[12] $$\sigma_U = \rho_U R/3 \qquad (38)$$

---

[12] It will be understood that geometrization of the corporal content of the cosmos as an area-density creates only a virtual ubiquitous impedance - the volumetric density ρ_U is still essentially zero.

As with the flat plane model shown in **Fig 4**, the effective inertial density of the spherical model of the cosmos is enhanced by a factor **R/3** as the geometry becomes Euclidean. Large shells can be simulated as flat surfaces - consequently they are functionally indistinguishable from infinite flat planes (**Fig 1**). While there is no physical change to the structure, there is an operational change that takes place as the universe expands from a small high density volume to its present size – the virtual energy difference between the 3-sphere and 2-sphere cosmic models takes into account the difference between mass assembled on a surface and mass distributed throughout a volume. That the universe acts as a ubiquitous virtual area density in opposing acceleration, it must dedicate its total mass-energy M_U amongst the three dimensions of space. Moreover, the real energy contained in a sphere of radius R assembled as a surface area is 5/6 that created by the same mass assembled as a volume. Hence, if U_2 is the energy of the 2-sphere shell universe and U_3 is the energy of the 3-sphere universe, both assembled from the same mass M_U, then:

$$U_2 = \frac{M_U^2 G}{2R_2} \qquad U_3 = \frac{3M_U^2 G}{5R_3} \qquad (39)$$

The difference in the two expressions is a result of the binding energy -- the action of gravity acting upon gravity. The modified inertial area-density U_2 will be the sum of the individual shell densities required to create the nested shell construct from the same Hubble mass M_U. Mass deficit in constructing the universe as a set of concentric shells equals the difference between the 2-sphere energy U_2 and the 3-sphere energy U_3. The same result follows as the universe expands -- a gradual shift in the retro-directive efficiency of inertial space could impact the expansion rate. In any event, the present state of the cosmos requires a size adjustment to account for the gravitational energy lost in transforming from a uniform density sphere to area density functionality. The energy available in both cases is M_U, hence the 2-sphere shell would need to have a smaller radius to produce the same internal **G** field, hence:

$$5R_3 = 6R_2 \qquad (40)$$

When the Hubble functions as a surface area (whether as a cumulative density or as a plurality of nested shells), it will behave functionally as a 2-sphere of radius $R = R_2$,

$$R = R_2 = (5/6)R_3 = (5/6)R_H \approx (0.83)(1.3 \times 10^{26}) \approx 1.08 \times 10^{26} \text{ meters} \qquad (41)$$

Fig 8: Inertial pressure **a[σ_U]** experienced within an accelerating body by the virtual area density **σ_U** compiled from the real area densities of planes both forward and aft over a length length **L**. Hence, therefrom, total virtual inertial area density **σ_U** must equal **[$\rho L/3$]**. For the Hubble universe, **L $\approx 10^{26}$** meters. Momentum flow is inward (opposite red arrow) and equal virtual inertial pressure created by the acceleration '**a**' upon the surface of **M**

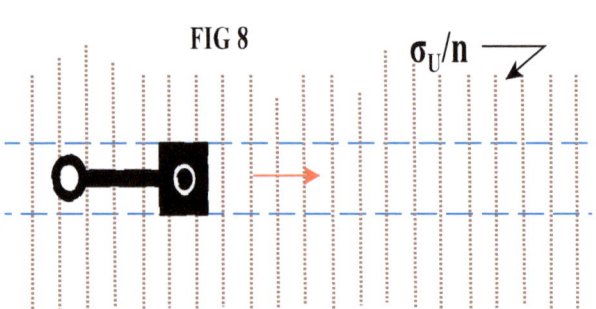

An accelerated piston (red arrow) encounters cosmic virtual inertia as retro directive area density from each laminate. This virtual inertia exists as the gravitational distension normal to each imagined plane into which the Hubble sphere could be sliced. Gravity fields of parallel infinite planes are additive, the intensity of the **g** field at every location depending upon the geometry of the mass that determines the pseudo force reactance. For infinite planes, the **g** fields thereof are perpendicular thereto. Consequently the operative inertial effect of each **σ_U/n** laminate exists throughout the volume defined by the dotted blue cylinder. For a single thick slab commensurate with the Hubble scale **R**, area density **σ_U** has the ubiquitous value **1 kg/m²** consistent with the **g** fields associated therewith. Although the **g** field of each laminate is canceled by the **g** field of an adjacent laminate, every **g** derives from an inertial mass. The readiness by which two adjacent inertial laminates respond to the call of an external force is cumulative not cancelling.

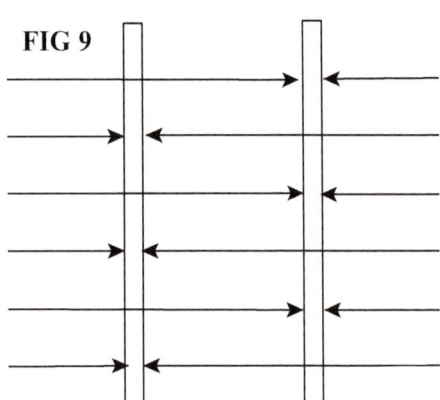

FIG 9

Fig 9 depicts a universe as two parallel plains separated by a distance **R**. The area of each plain is **$2\pi R^2$** and each contains ½ **M_U**, so the area density is **σ_U**. Arrows show the gravity field of each plane. Outside the planes the gravitational intensity is **$4\pi G\sigma_U$**, but in the space between the planes the gravi8ty intensity is zero, that is:

$$\sigma_U\left[\frac{c^2}{R}\right] = \sigma_U\left[\frac{c^2}{R}\right]$$

The **G** fields of each plane are equal and opposite. Thence, internal stress tension (negative pressure) equals:

$$-P = \left[\frac{\sigma_U c^2}{R}\right]$$

A body within feels no **g** force from the cosmos. In the two plane simile, the planes can be though of as opposite sides of the Hubble shell model. Alternatively, matter could be evenly distributed throughout the volume betwixt the planes to create a plenum of planes which act the same as the two parallel surfaces. The two planes represent the mass of the universe spread over the Hubble surface.

Propagation Velocity of Forces

In the traditional analysis of waves as vibrating particles, propagation speed is determined by the characteristics of the medium and the nature of the wave. Whether these methods can be applied to gravity and inertia, will depend first upon an understanding of space as a medium and secondly whether either gravity or inertial reaction can be understood to propagate in the conventional sense. That empty space is known to function has a medium for the prolongation of transverse EM waves, there is good reason to suspect it also functions as a medium for propagation of transverse gravitational waves. Indeed, if the pressure and density ratio are as is required for the maintenance of zero energy in an expanding null universe, then per ( )

$$-3P = \rho c^2$$

For a transverse wave in a medium having pressure **P** and density ρ from, the propagation speed v is:

$$v = \sqrt{\frac{3P}{\rho}} = \sqrt{\frac{-\rho c^2}{\rho}} = ic$$

In classic mechanics, the bulk modulus relating an increase in pressure to the fractional decrease in volume is symbolized as β:

$$\beta = \frac{\text{change in pressure}}{\text{fractional change in volume}} = (-)\frac{\Delta P}{\Delta V/V}$$

The reciprocal of the bulk modulus is the compressibility:

$$k = 1/\beta = -(1/P)[(\Delta V)/V_o)] \qquad (45)$$

where the (--) sign indicates that for real fluids and positive pressures, increased pressure results in smaller volume. An incompressible real fluid requires $\Delta V/V = 0$, then $\beta \longrightarrow \infty$. Whether a mass or fluid under negative pressure changes volume is not known. If negative internal pressure causes contraction, then a pressure increase from negative to less negative will cause expansion, in which case the bulk modulus is positive. If negative pressure causes no change in volume, then a pressure increase from negative to less negative will cause no change, in which case compressibility is zero. In classical theory, longitudinal propagation velocity v_L is:

$$v_L = \sqrt{\frac{\beta}{\rho_U}} \qquad (46)$$

Two curious possibilities ensue. The first is that of whether negative pressure space can be considered an incompressible fluid, in which case per (46), a longitudinal pressure wave travels at infinite velocity (any change in pressure is felt instantly throughout the Hubble volume). The 2nd arises from the fact density ρ is zero in empty space (notwithstanding a virtual area density σ_U at all locations), the real density is zero in empty space, hence v_L may be plausible infinite for more than one reason.

From (29) and (46)

$$(v_T)^2 = c^2 = (-3P/\rho_U) \qquad (52)$$

That (46) depends from (29), it qualifies as the equation of state for a constant "**c**" is universe:

$$\frac{\partial P}{\partial \rho} = \mathbf{0} \qquad (53)$$

The precedent for existence is encoded in the zero energy mandate. Positive $\mathbf{Mc^2}$ energy must be balanced by negative spatial energy. Thence, the ratio between pressure and density is constant (equal to $\mathbf{c^2}$). While there is academic interest in the plausibility of instantaneous action at a distance, the reality of space as a local inertial ether $\boldsymbol{\sigma}_U$ is the only cosmologically coherent solution[13] That resistance to acceleration is instantaneously experienced without the requirement of action at a distance, there is no need to further pursue force propagation alternatives.

---

[13] $\mathbf{c^2}$ as the panoptic constant of space *a la* (52). The intrinsic communion between space and time is manifest in the constancy thereof. That (47) derives from (29) and (46), the equation of state for the cosmos reduces to $(-\mathbf{c^2})$. An increase in longitudinal pressure does not result in a decrease in volume. Consequently $\mathbf{dP/d\rho}$ is **0**.

Virtual Impedance From negative Pressure

As Illustrated herein previously (**Fig 6**), the σ_U impedance is ubiquitous, hence superimposed upon an accelerating body at the point of action. No physical motion is involved, nor is there a traveling spatial parcel per se. The need for an incompressible ether is abrogated. The workings of the universe are marvelously economic, the impedance σ_U ever-present at all places.[14]

From Newton's 2nd law:
$$\mathbf{F} = \mathbf{M}_U \mathbf{a} = \rho_U \mathbf{V a} \tag{54}$$

Whence the Pressure:
$$\frac{F}{m^2} = P = \frac{[4/3][\pi R^3 \times \rho_U(a)]}{4\pi R^2} = \frac{\rho_U(a) R}{3} = \sigma_U(a) \tag{55}$$

For a point mass $\mathbf{M}_p$, then:
$$a = 4\pi M_p G \tag{56}$$

The '**g**' field is obtained by dividing by the area
$$g = \frac{4\pi M_p G}{4\pi r^2} = \frac{M_p G}{r^2} \tag{57}$$

From Friedmann's equation:
$$4\pi G \rho_U = 3qH^2 \tag{58}$$

Whence:
$$G = q\frac{3H^2}{4\pi \rho_U} = -\frac{3c^2}{4\pi R^2} \times \frac{4}{3}\left[\pi R^3\right]\frac{1}{M_U} = \frac{Rc^2}{M_U} \tag{59}$$

Therefore, when $\mathbf{r} = \mathbf{R}$ and $\mathbf{M}_p = \mathbf{M}_U$, the Hubble '**g**' field equals:
$$g_U = M_U G / r^2 = (M_U [c^2 R]/M_U) R^2 = c^2/R \tag{60}$$

In an early epic when $\mathbf{R} \longrightarrow \mathbf{0}$, expansion acceleration $\longrightarrow \infty$ for an infinitesimally short duration.

---

[14]The volume of a volume of empty space is unaffected by negative pressure. Consequently a decrease in the magnitude of negative pressure produces no change in volume, consequently there is no change in volume so long as spatial pressure $P < 0$. Bulk modulus $\beta \longrightarrow \infty$, consequently compressibility is zero. For unidirectional accelerations, Young's modulus $\mathbf{Y} \longrightarrow \infty$.

Adverting again the geometry(s) depicted in **Fig 2**. That the **g** field of each slab of blocks is added to the **g** field of every other parallel slab, the **g** field will be the same as though the universe acted as one thick infinite area plane (which it does). Inasmuch as the area density σ_U representing the one cosmic slab operative can be placed anywhere within the slab, it can be placed everywhere. That the cumulative '**g**' field depends upon the total area density of all slabs combined, likewise so also will the inertial property of the universe at any location, depend upon the area density of all slabs combined. Each addition to cosmic dimensional topology enhances the effectivity of the energy density as a reactionary impediment. Just as the earths internal '**g**' field increases linearly with distance from the center:

$$F = \frac{4\pi G \rho r}{3} \tag{61}$$

To apply (61) to the Hubble, $r = R$, $\rho = \rho_U$, and in lieu of volumetric acceleration ($4\pi G$), the generic expression for the inertial impedance of the cosmos reduces to pressure per unit of acceleration:

$$\frac{F/A}{a} = \frac{\rho_U R}{3} = \sigma_U \tag{62}$$

which is identical with (55). Thus while the universe imposes an impedance σ_U upon every element of mass undergoing unidirectional acceleration, that is not the case with gravity. The gravity field of a body depends upon momentum inflow, that crossing the surface that encompasses the mass

Using the well understood physics of spheres, Newton's Law of gravity is revealed as a special case of his second law of motion. Although the proposition for local '**g**' fields as pseudo force distortions of the global **G** field has been previously proposed, it now becomes compelling:[15] For a uniform density sphere of mass **M** and radius '**r**,' Newton's 2nd law for surface pressure can be written in terms of ntn per meter^2:

$$\frac{F}{A} = \frac{F}{m^2} = \frac{Ma}{4\pi r^2} = \frac{c^2}{R} \times \frac{M}{4\pi r^2} \tag{63}$$

Newton's gravity law for the same sphere can be written as:

$$a = \frac{F}{kg} = \frac{MG}{r^2} \tag{64}$$

---

[15]The cosmos as a whole must account for inertial reaction fields (spatial pseudo forces) provoked by non-expanding forms of mass. In a perfectly uniform universe, there would be no gradients and therefore no gravity. It is to the lumpy universe, we owe our existence. Matter distorts spatial expansion. The perception of pressure over the earths surface is momentum influx. Momentum flow is convergent, total influx at any distance from the surface of a uniform spherical mass is constant - that it is distributed over a larger area accounts for inverse square diminution in the force gradient with distance.

Dividing (63) by (64)
$$\frac{\frac{F}{m^2}}{\frac{F}{kg}} = \frac{kg}{m^2} = \frac{c^2}{R(4\pi r^2)} \times \frac{r^2}{MG} \qquad (65)$$

Then
$$G\frac{kg}{m^2} = \frac{c^2}{4\pi R} \qquad (66)$$

From which:
$$4\pi G\left[\frac{kg}{m^2}\right] = \frac{c^2}{R} \qquad (67)$$

Thence:
$$G = c^2/4\pi R\sigma_U \qquad (68)$$

From (67) and (68), **G** and **ΛR** can now be understood as alternative expressions for spatial exponentiation, differing numerically only by **4π** [a natural eventuate of Newton's formulation which considers ($1/r^2$) diminution as separation distance rather than a ($1/4\pi r^2$) field effect]. Big **G** also encodes the [$1/\sigma_U$] factor in the dimensionality [$m^3/sec^2/kg$] = [m/sec^2]/[kg/m^2].

The derivation of **G** as a parameterized Hubble functionality, requires positive and negative energy be created in equal amounts. Net energy is zero, therefore average pressure is negative. Negative pressure induces expansion, expansion of negative pressure volume creates positive energy in form of enhanced inertia **M**. From Friedmann's equation:

$$4\pi G\rho_U = 3qH^2 \qquad (69)$$

For an exponentially expanding Hubble sphere, $q = -1$, and $\rho_U = M_U/(4/3)\rho_U R^3$, $H = c/R$, hence:

$$G = \frac{Rc^2}{M_U} \qquad (70)$$

And from (68):
$$G = \frac{c^2 R}{4\pi R^2 \sigma_U} \qquad (71)$$

Comparison of (70) and (71), then:
$$M_U = 4\pi R^2 \sigma_U \qquad (72)$$

From (72), cosmic mass increases in proportion to Hubble area. Consequently, cosmic density diminishes as:

$$\rho_U \propto 1/R. \qquad (73)$$

Appendix A - Transformation from Hubble Sphere Infinite planes

Transforming the Hubble to a 2-sphere shell density having the same mass requires a radial adjustment to the area over which the mass is spread.. To represent the inertial field of the same mass as three orthogonal flat plane area densities, the following relationships apply. .

Hubble Sphere Radius $R_H \approx 1.3 \times 10^{26}$ m

Gravitational energy $= U = \dfrac{3M_U^2 G}{5R_H}$

Hubble Shell Radius $R_2 = 1.08 \times 10^{26}$ m

Gravitational energy $= U = \dfrac{M_u^2 G}{2R_2}$

Surface area density $\sigma_U = M_u / 4\pi R_2^2$

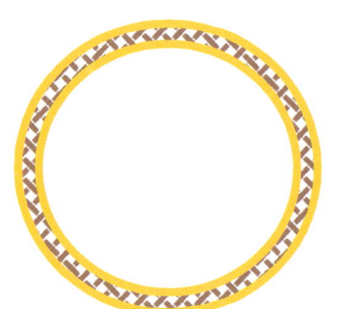

$\sigma_U = (\rho R_2 / 3)$,

Energy for the 3 planes $= (3\sigma_U)(4\pi R_2^2)$

$E_T = 3(\rho R_2/3)(4\pi R_2^2) = \rho_2 V_2 = M_U$

The operative inertial characteristics of the Hubble are defined by three orthogonal planes each having area density σ_U = one kg/meter squared.

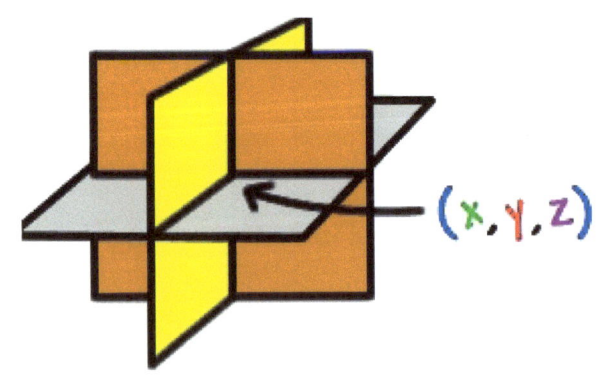

Appendix B - Circulatory Space Resolves the Dark Matter Riddle

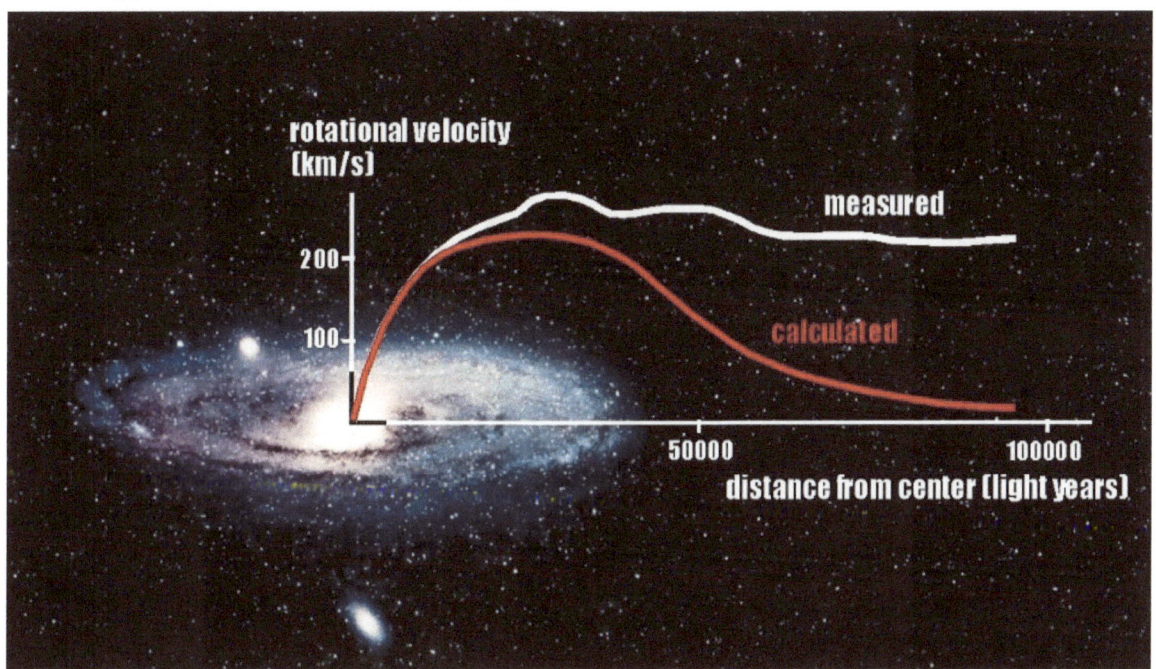

It has been more than three Centuries since D'Alembert proclaimed perfect fluids exert no pressure drag when moving at constant velocity. A mystery at the time, but now understood as a mandate of frictionless in-viscid fluids. Objects move through space at constant velocity unopposed. But acceleration of space wrt mass or vice versa, creates Newtonian forces.

Fig A-2 depicts concentric stream tubes of rotating space each having velocity $v = r\omega$. The continuity equation for an ideal incompressible circulating fluid (where density is constant) is:

$$A_1 v_1 = A_2 v_2 = Q \qquad (A\text{-}1)$$

A solid disc rotating at constant angular velocity $\acute{\omega}$, experiences centripetal force $\mathbf{F} = r\omega^2$ at any distance r. For circulatory space rotating as a solid disk, inward spatial flux is likewise proportional to r. To maintain orbital speed and distance, a star co-moving with the stream-tube flux requires centripetally directed spatial force $v^2 r(\sigma_U) = \sigma_U r \omega^2$ proportional to its orbital radius r (red arrows). In words, the centripetal force required to explain the stellar velocities of rotating galaxies is provided by co-rotating spatial flux having a characteristic inertia σ_U. A star at 'r' experiences inward pressure proportional to r due to circulatory space, Ergo, star velocities will be less dependent upon r since:

$$(v^2/r) \times r = v^2 .$$

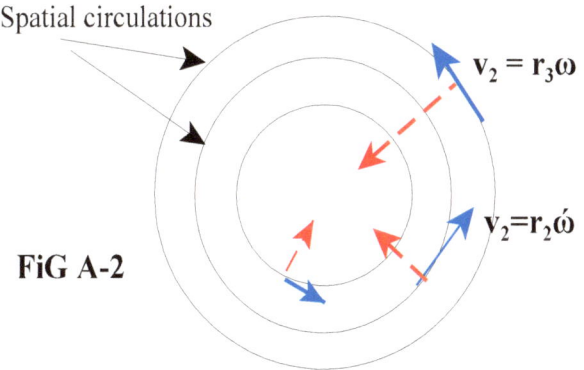

FiG A-2

Appendix C: Exponential Expansion Resolves the Creation-Inflation Paradigm

The prevailing state of cosmogony reckons the universe in terms of initial conditions. But what determined the initial conditions. The riddle of origin turns upon the present state of the cosmos. But according to current thinking, cosmological evolution takes form as a sequence of transitions to states of progressively lower symmetry. An unexplained temporal beginning followed straightaway by an *ad hoc* inflationary process of rapid spatial flattening growth coincidentally resolving the horizon problem. Interesting if true, but convenient otherwise. All of which raises the issue of a temporal beginning and how can the present be explained in terms of the past? To eliminate genesis, it is necessary to regress the present to a potentiality that feeds upon change.

For present purposes, the observable universe will be taken as the Hubble sphere centered on the observer. Parameterization of the expansion function takes form as Hubble velocity squared divided by Hubble scale:

$$\ddot{R} = \frac{\dot{R}^2}{R} \quad \text{-----> } (-q) = \frac{\ddot{R} R}{\dot{R}^2} \tag{C-1}$$

Where (q) is the deceleration parameter [having a value of (-1) for an accelerating universe].

At the Hubble limit, (C-1) reduces to:

$$\ddot{R} = a_n = \frac{c^2}{R} \tag{C-2}$$

Where a_n denotes the present value $\ddot{R}$. Taking (C-2) as the defining operative, the need for imposition of initial conditions is disposed. As the imaginary cosmic clock is run backwards, $R \longrightarrow 0$. The "now" rate of spatial expansion extrapolates to a state of infinite acceleration for an infinitesimally short duration. The potential for a change of state from non-existence to runaway expansion intrinsically pervades, cogent, compelling and eventually overpowering. Once set in motion, it cannot be reversed - the universe is a self sustaining positive feedback organic. The greater the volume, the greater that rate of volumetric growth. An existence that is known to expand exponentially if initiated, bespeaks of a subliminal potentiality. Expansion appears to be unavoidably, the universe inevitable.

Spatial exponentiation is a natural result of the zero energy universe. Negative energy and positive energy must be created in equal amounts during expansion. From Friedmann's 2nd equation:

$$\ddot{R} = -\frac{4\pi G}{3}\left[\rho_U + \frac{3P}{c^2}\right]R + \frac{\Lambda R}{3} \tag{C-3}$$

When pressure P is negative and equal $\rho_U c^2/3$, then (C-3) reduces to:

$$\ddot{R} = \frac{\Lambda R}{3} = \frac{3H^2 R}{3} = \frac{c^2}{R} \tag{C-4}$$

Which has the solution:
$$R = e^{Ht} \tag{C-5}$$

Creation *Ex Nihilo* is a natural consequence of the initial condition ($t = 0$). Acceleration is infinite for an infinitesimal instant, the universe is geometrically flat and there are no horizon limitations.

Appendix-D: The Null Universe Expands Exponentially Without Dark Energy

In his last draft of the General Theory of Relativity, Einstein introduced a cosmological constant lambda {Λ} believed necessary to prevent his static model of the universe from collapsing due to the inward pull of gravity. Within a decade, however, the evidence for expansion had reached a compelling level. Einstein was remorse. By adding Λ as a balancing factor, he had deluded himself into a false sense of theoretical *compléter*. [Indeed, it is likely that Einstein would have resolved the ostensible problem of gravitational collapse by inventing some sort of global dynamic]. And that dynamic might have been congruent with what we now know about the expansion of the universe

We need not, however, speculate further about what Einstein might have accomplish had he guessed differently about the static state of the universe. But alas, that was not to be -- or was it? Much has been learned in the years since the discovery of Expansion. The irony of the cc could not be appreciated until the last few years of the 20t h Century... finally made manifest in the latter years thereof. In retrospect, it is irony of the cosmological constant, Einstein not only predicted the expansion of the universe, he calculated the acceleration rate.

To balance gravity, Einstein's cosmological constant Λ must equal $3H^2$ [per (C-4)]. But (C-4) is identical to (C-2). Ergo, Einstein's cosmological constant is identical to the cosmological acceleration parameter

$$\ddot{R} = \frac{c^2}{R} = \frac{\Lambda R}{3} \tag{D-1}$$

In Einstein's formulation, to cancel gravity, the Λ factor needs to have a value $4\pi G\rho$. Consequently G can be expressed as

$$G = \frac{\Lambda}{4\pi\rho_U} \tag{D-2}$$

Substituting $M_U/[(4/3)(\pi R^3)]$ for ρ_U and Einstein's value $3H^2$ for Λ:

$$G = \frac{Rc^2}{M_U} \tag{D-3}$$

Per (72),

$$G = \frac{Rc^2}{4\pi R^2 \sigma_U} = \frac{c^2}{R} \times \frac{1}{4\pi\sigma_U} = \frac{\ddot{R}}{4\pi\sigma_U} \tag{D-4}$$

Einstein searched for a cosmological acceleration factor to balance **G**. But alas, the acceleration factor ΛR discovered by Einstein, is not a separate field, but rather the dynamic essence of the expansion field that drives the universe. In misinterpreting **G** as a principal constitutive of static space, Einstein deduced the need for balancing factor. The reality of expanding space is that **Big G** is divergent. The little 'g' fields of individual masses are emergent pseudo forces.

Appendix E - The Inertia of Gravity - 'g' field energy as inertia.

When a moving object accelerates by changing velocity reactionary force is instantaneous. The dependence thereof upon the global extent of an objects '**g**' field augments the already extensively developed proposition founded upon infinite plane dynamics. Mass energy in form as the '**g**' field of a local body is congruent with infinite plane area density theory as well 3-D volumetric density (gravitational intensity falls off inverse squared, so the force integral over any sphere concentric with the mass will be the same at all radii).

Gravity fields and inertial reactions depend from relative acceleration between space and mass. That inertial reactions and gravity both depend from the **Ma** product, there is good reason to consider gravitation as something other than a propagation phenomena. The dependency of instantaneous inertial reaction upon the **g** field of a mass follows from the proposition that gravity and inertia are cosmological in origin. Local objects distort the expansion field **G**. The '**g**' field of a local body is but the instantaneous inertial reaction (pseudo force) created by inertial resistance of mass-energy to isotropic spatial expansion The action of **G** upon **M** is predicate —> Mc^2 mass energy follows —> from which the inertia of the non-expanding volume occupied by the Mc^2 energy, imposes a counter acceleration field '**g**' that appears to miraculously emanate from within.

To evaluate the gravitational effect of a given quantity of bare mass M_B, it can be imagined ground into small grains uniformly spread of the volume of the Hubble universe. The pressure **P** on the Hubble surface can then be considered as though all the particles are radially accelerated to create an area density surface pressure:

$$P = \frac{c^2}{R} \times \frac{M_B}{4\pi R^2} \tag{E-1}$$

Energy of a volume **V** of pressure **P** is:

$$E = 3PV = 3\frac{c^2}{R}\left[\frac{M_B}{4\pi R^2}\right] \times \left[\frac{4}{3}\pi R^3\right] = M_B c^2 \tag{E-2}$$

As was the case with cosmic mass $M_U c^2$, the $M_B c^2$ energy of individual masses owe their Mc^2 energy to the wakeless expansion factor c^2/R embedded in the formalization of **G**. Spatial expansion powers the cosmos. If M_B were instead, compacted into a high density flat disk of radius '**r**', the '**g**' field thereof would be confined to a cylindrical tube of length **R** and area $4\pi r^2$ (analogous to that created by an accelerated. Piston (e.g., Fig 8). From (10), the '**g**' field pressure can be expressed in terms of '**g**' force on the cylinder ends (the cylinder connects the flat disk representing the earth with the infinite plane slices that comprise the universe. Hence, the **g** field lines between σ_B and σ_U are parallel.

$$\sigma_U g = \frac{\sigma_B c^2}{R} = \frac{M_B c^2}{R(4\pi r^2)} = P \tag{E-3}$$

For parallel **g** lines, $E = PV$:

$$E = PV = \frac{M_B c^2}{R(4\pi r^2)} \times (R)(4\pi r^2) = M_B c^2 \tag{E-4}$$

Appendix F - Spatial Inertia Explains 'g' Fields as Pseudo Forces

A local gravity field can be cancelled by transformation. A free falling body in a uniform gravitational field, experiences no 'g' forces. While weightless condition can be approximated by a large spherical mass, complete abrogation of gravitational influence exists only in the form of an infinite plane of uniform density. **Fig F-1** simulates the Hubble universe as two halves which for inertial purposes can be considered as two thick parallel planes accelerating apart (green arrows) creating equal and opposite **g** fields (red arrows) upon a cube (blue) between. Neglecting all forces other than those parallel to the **X** axis, the **g** field entering the left face of M_B equals the **g** field entering the right face of M_B. From Gauss's law of gravity, each **g** field equals $2\pi\sigma_U G$.

The fact that the left and right pointing flux lines are of finite strength, reveals the universe is finite in operative scope.[16] The directional acceleration factor **α** for both halves is:

$$\alpha = 2 \times \sigma_U(2\pi G) = (\sigma_U)(4\pi G) = kg/m^2 (12.56)(6.67 \times 10^{-11} \, m/sec^2) = c^2/R$$

Where the operative value of **R** is not 1.3×10^{26} meters but rather 1.08×10^{26} meters. In the infinite plane model of the universe, isotropic Hubble acceleration is broken down into three coordinate directions **X, Y, Z** wherein only **X** need be considered to explain the conceptual interplay between inertia and gravity. For example, one can calculate the gravity field of the a spherical mass of uniform density (here approximated as the earth where $M_E = 5.98 \times 10^{24}$ kg and $r_E = 6.37 \times 10^6$ meters by considering the earth as squashed into a cube. Then the directional acceleration is:

hence:
$$g\sigma_U = \alpha\sigma_E = \sigma_U(4\pi)G\sigma_E$$

$$g = 4\pi G\sigma_E = \alpha\sigma_E$$

FIG F1: G field spatial momentum divergence (green arrows) are impeded by any form of inertial mass energy M_B contained within the non-expanding cube having face area **A** (blue). The difference in the expansion rate of a volume containing Mc^2 inertial energy and an empty volume, determines the pressure difference **P** on the surface area **A**:

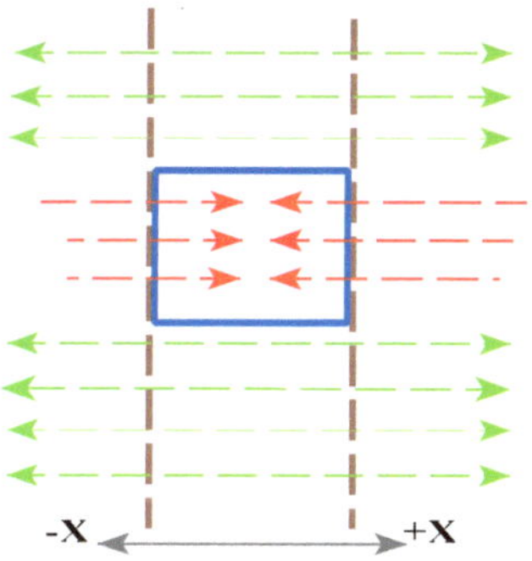

$$-- P = (\alpha)[M_B/A]$$

Application of c^2/R acceleration to M_B in the +X direction, reduces the **g** field on the right face to zero while doubling the apparent **g** field on the left face. Total momentum influx remains the same. There is thus no difference between momentum flow pressure created by spatial acceleration wrt mass and mass accelerated wrt space.

---

[16] If this were not the case, even though cosmic density is low (in the range of 10^{-26} kg/m³), a universe of unlimited thickness in the **X** dimension would result in infinite **g** field

Appendix G. The Convergence Model of Gravity as a Pseudo Force

Pseudo forces arise from accelerations. An accelerated body in free space experiences internal counter forces proportionate to the inertial mass. As Richard Feynman observed, gravitational forces, are like pseudo forces in this regard, they are always proportion to the masses.[17] Would a mass **M** at rest in an accelerating universe, also experience an internal reactionary force? This is Einstein's idea of relative acceleration, it should be impossible to detect whether **M** or the universe is accelerating.

Standard physics has no explanation as to how a counter force is delivered to the mass of an accelerated body in free space (far removed from the influence of other fields). Even though no part of the physical universe is in contact therewith, inertial reaction is a fact. The answer comes from the large scale structure of the universe. The cosmos speaks not as rarified vacuum comprised of widely separated clumps of matter, but rather as a local infinite plane of uniform density.

Originally conjured by Einstein in 1917 as an ad hoc factor to prevent gravitational collapse of his static model of the universe, the cosmological constant Λ, can now be understood as an embedded factor of the global acceleration parameter **G**, specifically, using Einstein's value $\Lambda = 3H^2$

$$4\pi G\sigma_U = \Lambda R/3 = 3H^2 R/3 = H^2 R = c^2/R \tag{G-1}$$

Whence:

$$G = \frac{\Lambda R}{3} \times \frac{1}{4\pi\sigma_U} = \frac{H^2 R}{4\pi\sigma_U} \tag{G-2}$$

While $\Lambda R/3$ is no longer tasked with balancing gravity, it does deploy as the global expansion field from which 'g' forces emerge. When multiplied by the mass **M** of a non accelerating body, it specifies the gravitational counter force. For example, to find the 'g' field of a spherical mass of radius 'r', one simply calculates total inertial reactionary acceleration from Gauss's Law of gravity:

$$\oint \mathbf{g} \cdot \mathbf{ds} = [M \times 4(\pi)G] \tag{G-3}$$

Then divide by integral of **ds** over the surface area of the sphere ($4\pi r2$) to obtain the intensity (acceleration per unit area):

$$g = 4(\pi)MG/4(\pi)r^2 = MG/r^2 \tag{G-4}$$

For a spherically uniform mass, Newton's law of Gravity (G-4) can be reckoned as isotropic counter acceleration, opposite to **G**, proportional to **M** and convergent thereon (the **F** = **M**a inertial reaction of **M** opposing isotropic spatial acceleration **G**). Per (G-2), Newton's gravitational constant **G** is seen to be but an alternative expression for the cosmological expansion rate **[(c^2/R) per kg]**. Ironically, this is one-in-the-same as that derived a century past by Einstein [$\Lambda R/3$] to balance gravity (eventually validated *a la* the 1998 supernova studies). Viewed herein, however, neither as a factor to balance gravity, nor as a separate force *per se*, but rather as the cause of gravity.

---

[17]"One very important feature of pseudo forces is that they are always proportional to the masses. The possibility exists therefore, that gravity is a pseudo force."[Feynman, lectures on Physics]

Appendix H - An Angular Momentum Model of the Electron

To reckon circulatory space as a particle *completers*, it must portray concurrently as angular momentum and inverse squared radial force. As a quantum mechanical concept, spin is considered metaphorical, but herein classically portrayed as three dimensional spatial vorticity. Quantum mechanical spin can then be analogized to gyroscopic precession in that the three orthogonal planes of rotation are linked by the laws of conservation. A 360 degree reorientation of one spatial spin plane results in a 180 degree shift in the two orthogonal spin planes.[18] By contrast, quantum spin states are described by vector-like objects called spinors.[19] Rotating a half spin (**h/4π**) particle by 360 degrees does not bring it back to the same quantum state. To return the particle to its original state, one needs a 720-degree rotation.

To actually derive the strength of the electric field from the physical properties of the electron, spin must a corporal reality, and angular momentum must involve tangible circulatory motion. As with all forces based upon relative action between matter and space, only motions involving acceleration are detectable. Circulatory space involves continuous centripetal acceleration. The foremost question simplifies to the causal source of spatial circulation. In terrestrial vortices, low pressure is the primary culprit. Could low spatial pressure be a cause-in-fact of circulatory spatial flux and the consequent enigma(s) linked with that electron (the electric field and the isotropic angular momentum field).[20]

As a theory, captivation of 3-D spatial vortices within regions of intense negative pressure is speculative. While it is well established that a photon in the vicinity of a nucleus can transform into an electron-positron pair, the degree to which negative energy in the near field of the nucleon plays a part is not known. For present purposes, matter is provisionally considered as a negative pressure sink hole characterized by an external circulatory spatial field. In the case of electrons and positrons, the angular momentum electric field constitutes the essence of the construct.

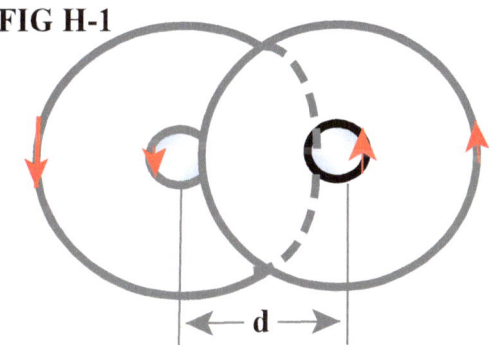

FIG H-1

Fig H-1: Spatial circulation explains the electrical and mechanical properties of the fictitious electron charge **q**. Two electrons each composed of circulatory space within a volume of negative pressure **-P** communicate through their respective extended spatial circulations, opposing in the near space between centers and reinforcing in the hinter land beyond. Weakening the near field and augmenting the far field, creates circulatory force imbalance. Like charges repel.

---

[18] Stable particles such as electrons and positrons exhibit isotropic spatial angular momentums **h/4π** which account for their 720 degree spin symmetry (One full 360° rotation flips the other two (orthogonal) angular momentum planes each 180 degrees. For a spatial circulation as opposed to a mechanical rotation, one can envision a three dimensional kind of "torque" on an electron.

[19] A spinor is essentially an object which turns into its negative when it undergoes a 2π rotation.

[20] Richard Feynman: *"The most shocking and disturbing thing about quantum mechanics, is that if you take the angular momentum of an electron along any particular axis, it is always ℏ/2"*.

As to be developed, the defining properties of electrons as the fundamental unit of charge **q**, are mechanical in nature (mass m_o, energy radius r_o, angular momentum $h/4\pi$, and c^2). As with gravity, the abstract notion of spatial acceleration reveals as force-in-fact. At root, the symmetrical, aspect of Newton's 2nd law — irrespective of whether space or mass accelerates, angular momentum can be defined as circulatory space relative to a fixed or counter rotating center of mass. The spatial angular momentum field is **3-d** and global in extent, a new light upon an old paradox.[21]

The circulatory space model of the electron derives either as an ideal vortex [**Fig H-2** velocity increases as **1/r** to a maximum '**c**' at r_o] or as a constant **c** velocity circulation at all radii [**Fig H-3**] Both models produce the same total angular momentum. Based upon the classical kinship between vortical velocity **v** and radius **r**, taken together with the presumption of **c** as the limiting velocity at r_o, the following formulation ensues:

$$\mathbf{v} \cdot \mathbf{d} = \mathbf{c} \cdot \mathbf{r_o} \qquad \text{(H-1)}$$

The balanced centripetal force c^2/r_o due to the electrons own spatial circulatory field at r_o will be moderated by the field of a 2nd electron at distance **d** having a vortical velocity **v** per (H-1) as shown in **Fig H-I**. Whence net centripetal force at r_o is determined by the velocity v_2 of the 2nd electron at distance **d**, therefore:

$$\frac{F_d}{m_o} = \frac{v^2}{r_o} = \frac{1}{r_o}\left[\frac{c \cdot r_o}{d}\right]^2 = \frac{c^2 r_o}{d^2} \qquad \text{(H-2)}$$

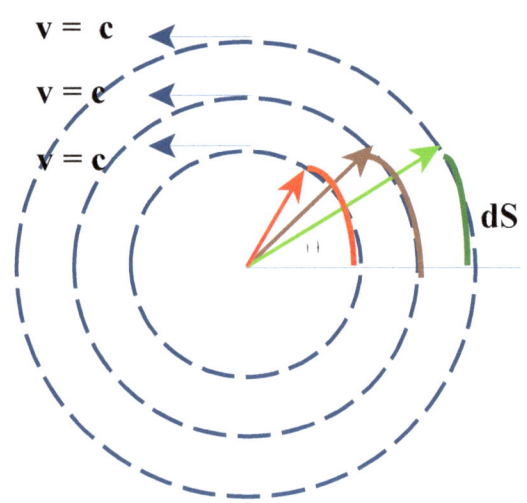

The same result can be arrived at from an angular momentum perspective by considering circulatory flux equal to **c** at all radii as shown in **Fig H-3**. The circulations sweep out the same distance **dS** per unit of time **dt** [ds is red for a small radius, green for a greater radius and brown for the largest radius]. The affect of a circulation at radius **r** to the angular velocity 'ω' = $d\theta/dt$ = $ds/r(dt)$ = c/r. The centripetal force due to the second electron c^2/d is spread over the circulation path $2\pi r_o$ divided by $2\pi d$. Hence:

$$F_w = \frac{m_o c^2}{d} \times \frac{2\pi r_o}{2\pi d} = \frac{m_o c^2 r_o}{d^2} \qquad \text{(H-3)}$$

Force $\mathbf{F_d}$ obtained from vortical theory and force $\mathbf{F_\omega}$ from angular momentum considerations, are identical. For electrons and positrons, circulations are 3-D. As a adjunct of both models, the electron size is illusory spherical shell of radius $\mathbf{r_o}$ determined by the energy required to deposit a charge $\mathbf{q}$ on the surface.[21] A shell of radius $\mathbf{r_o}$ having charge $\mathbf{q}$ has $\mathbf{m_o c^2}$ energy:

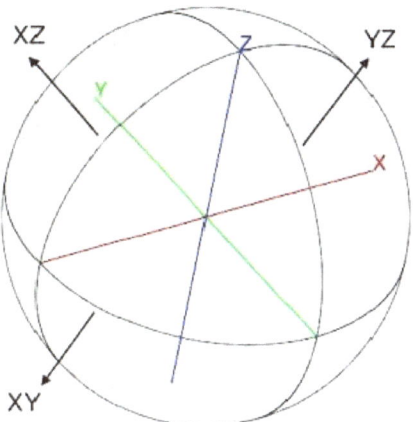

$$\mathbf{m_o c^2} = \frac{1}{4\pi\varepsilon_o} \times \frac{1}{2} \times \frac{\mathbf{q}^2}{\mathbf{r_o}} \quad \text{(H-4)}$$

from which

$$\mathbf{r_o = 1.41 \text{ fermi}}$$

Fig H-4: There are always two orthogonal circulatory planes passing through the centers of two communicating electrons, consequently total force is twice that calculated for a single plane of circulation. A 2nd electron on the **X** axis (red) of a 1st electron at the intersection of the red, green and blue axis, would experience circulations in the **XY** plane and **XZ** plane, but not the **YZ** plane.

The formulations (H-5) and (H-6) show the force produced by the components of circulatory space in the XZ and YZ planes equals the force obtained using Coulomb's law. That the two formalisms are force-wise equivalent, it will prove convenient thenceforth to envision charge as circulatory space.

$$F_e = \frac{k_e q_e^2}{d^2}$$
$$= \frac{(9 \times 10^9 \, kgmm^2/coul^2)(1.6 \times 10^{-19} \, coul)^2}{d^2}$$
$$= \frac{23 \times 10^{-29} \, ntnm^2}{d^2} \quad \text{(H-5)}$$

$$F_r = \frac{2 m_o r_o c^2}{d^2}$$
$$= \frac{(9.1 \times 10^{-31} \, kgm)(1.4 \times 10^{-15} \, m)(3 \times 10^8 \, m/\sec)^2}{d^2} \quad \text{(H-6)}$$
$$= \frac{23 \times 10^{-29} \, ntnm^2}{d^2}$$

---

[21] The Hub radius [also called the energy radius $\mathbf{r_o}$ (or charge radius in some literature)], is obtained from electron scattering experiments. That the results do not converge upon a certainty of size, can be considered indicative of the dynamic interplay between $\mathbf{m_o}$ and $\mathbf{r_o}$.

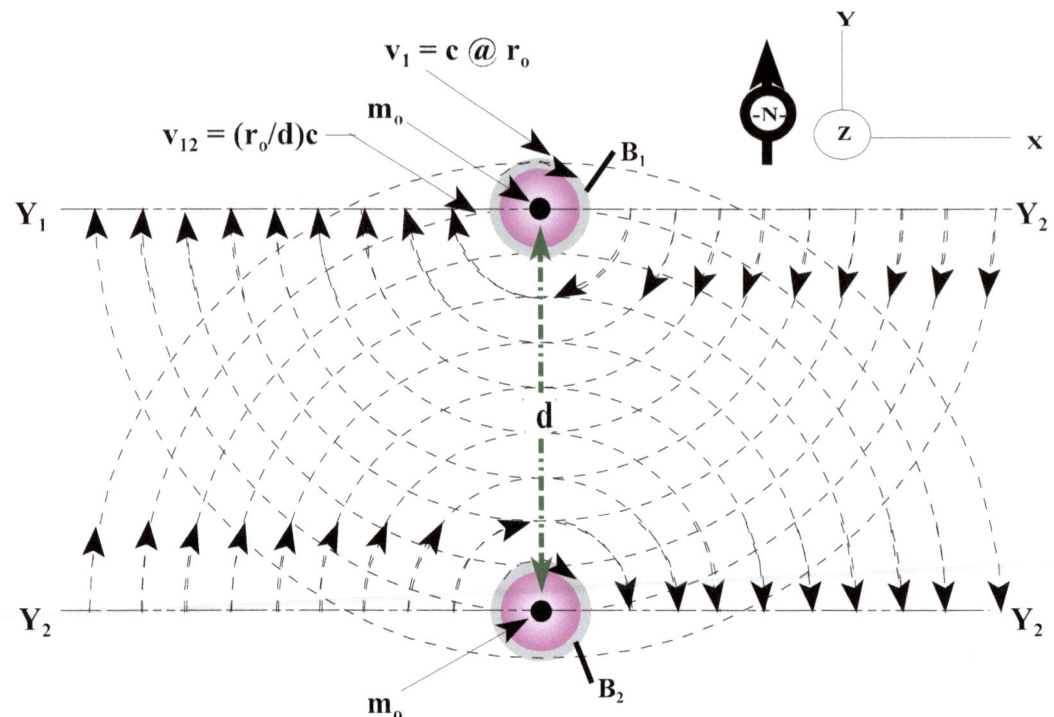

Figure H-5. The coupling between vortical fields is shown as being simultaneously both additive and subtractive. The two spatial rotations in the x-y plane each have peripheral spin velocity "c" at radius r_o, and each encompasses an identical symbolic mass m_o. Both circulations are clockwise in the **X/Y** plane which is divided into three areas by the two parallel east-west lines Y_1-Y_1 and Y_2-Y_2. In the region between these lines the vortical field of each particle counteracts the local angular momentum of the other. In the space north of Y_1-Y_1 the field of β_2 is depicted as augmenting the vortical strength of β_1 and in the area south of Y_2-Y_2 the field of β_1 is depicted as bolstering the strength of β_2. Superposition of the two fields results in an unbalance in the force exerted upon each of the masses m_o. The net force upon β_2 will be southward, and that upon B_1 will be northward. Like charges repel.

The $m_o c^2$ energy involved in building an electron by depositing bits of charge carried from infinity leads to the following:

$$E = \frac{k_e q^2}{2r_o} = m_o c^2 \tag{H-7}$$

From which the force **F** is:

$$F = \frac{k_e q^2}{d^2} = \frac{2[m_o c^2 r_o]}{d^2} \tag{H-8}$$

Which is (H-2), (H-3), (H-5) and (H-6).

Appendix J. The physical meaning of Einstein's Cosmological Constant Λ

Following the discovery of expansion, Einstein viewed the cosmological constant as a mistake, suggesting it be 'awayed." If the universe is expanding, reasoned Einstein, there was no need for a cosmological balancing force. Unappreciated at the time, the expanding universe had a balancing problem of its it own. The question would soon arise: How is it, after 13 billion years, gravity has remained in equilibrium with expansion? Much of the 20th Century was spend speculating upon whether gravity or expansion would ultimately prevail. Taking the universe as a giant sphere of mass M_U, perfect balance requires gravitational force equal the cosmological constant force $\Lambda R/3$:

$$F_G - F_\Lambda = -- GM_U/r^2 + \Lambda R/3 = -- 4\pi G\rho R/3 + \Lambda R/3 = 0 \qquad (J\text{-}1)$$

Hence:
$$\Lambda = 4\pi G\rho_U \qquad (J\text{-}2)$$

Volumetric growth of space $\dot{V}$ within the Hubble sphere and its derivative $\ddot{V}$ (volumetric acceleration) can be related to isotropic spatial flux $\dot{R}$ and its rate of change $\ddot{R}$. To find the production rate of space, we surround the Hubble with an imaginary Gaussian sphere S of radius R_S shown J-1. Accordingly, the following relations hold:

$$V = \frac{4}{3}\pi R^3 \quad\dotfill(J-3)$$
$$\dot{V} = (4\pi R^2)(\dot{R})$$
$$\ddot{V} = 8\pi R(\dot{R})^2 + 4\pi R^2(\ddot{R}) \quad\dotfill(J-4)$$

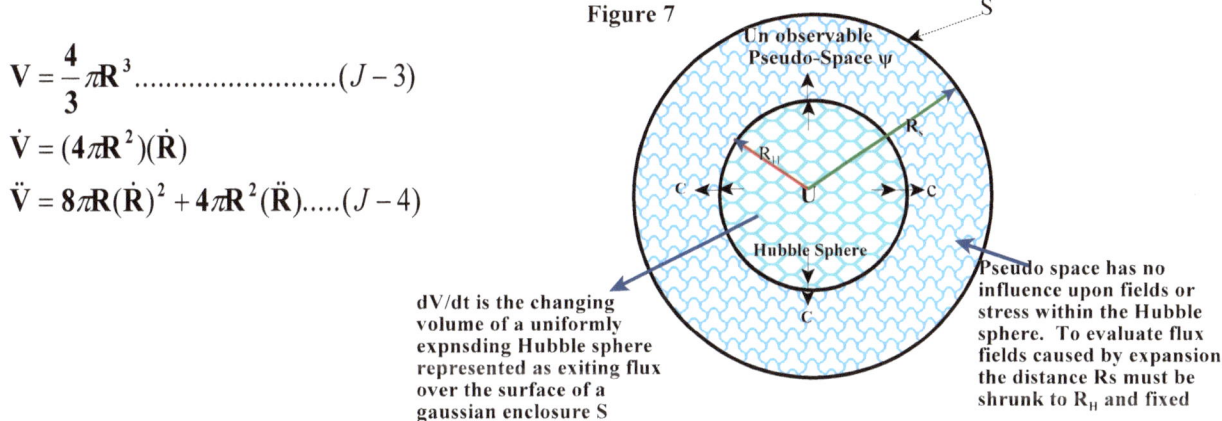

Figure 7

dV/dt is the changing volume of a uniformly expnsding Hubble sphere represented as exiting flux over the surface of a gaussian enclosure S

Pseudo space has no influence upon fields or stress within the Hubble sphere. To evaluate flux fields caused by expansion, the distance Rs must be shrunk to R_H and fixed

At the momentum of coincidence of the Hubble sphere with S, volumetric acceleration per unit area ($3H^2R$) precisely corresponds to Einstein's prescription for the cosmological counter force $\Lambda R/3$. But alas, G an Λ are not competing operatives vying for control of the universe. Rather Λ is embedded within G. From Gauss's Divergence theorem, *the integral of the volume containing the divergence is equal to the flux over the surface that contains the volume,* Hence from (J-4):

$$\frac{\ddot{V}}{A} = \frac{12\pi R\ddot{R}}{4\pi R^2} = \frac{3\ddot{R}}{R} = \frac{3c^2}{R} = 3H^2R = \Lambda R \qquad (J\text{-}5)$$

Einstein's cosmological counter force ($3H^2R$), now becomes rich in revelation. The existence thereof in form as isotropic spatial expansion, brings forth the issue as to the '*effect*' thereof upon inertial matter. The cosmological constant matches exponential expansion if $\Lambda = 3H^2$. The signature feature of GR, spatial curvature, is redundant. The $\Lambda R/3$ factor does not balance **G**, it is essence of **G**.

$$G = \frac{\Lambda}{4\pi\rho_U} \tag{J-5}$$

Therefore:
$$G = \frac{\Lambda R}{4\pi(3\sigma_U)} = \frac{3H^2R}{4\pi(3\sigma_U)} = \frac{H^2R}{4\pi\sigma_U} = \frac{c^2}{4\pi R\sigma_U} \tag{J-6}$$

The dimensionality of **G** (vol acceleration per unit mass) is a giveaway to its origin. The ratio (J-7): has been long puzzled over.[22]

$$GM_u/Rc^2 = 1 \tag{J-7}$$

However, it will be understood that (J-7) can be derived from Friedmann's equation by substituting $[M_U/(4/3)\pi R^3]$ for ρ_U. Simply put, (J-7) is an alternative expression for **G** in terms of Hubble parameters. M_U must equal $4\pi R^2\sigma_U$ and from Friedmann's equation, we derive (J-8): whence:

$$G = \frac{3H^2}{4\pi\rho_U} = \frac{3H^2}{4\pi[M_U/(4/3)\pi R^3]} = \frac{c^2 R}{M_u} = \frac{c^2}{4\pi R\sigma_u} \tag{J-8}$$

which is the same as (J-6). Expressing **G** in terms of a Hubble parameters reduces to spatial exponentiation divided by Hubble mass (J-9). Consentient therewith, '**g**' fields are exposed as nothing more than reactionary pseudo forces. The once exalted status of gravity as a primary fundament with its own special constant **G**, can now be relegated to an emergent consequence. The root cause of gravity is subsumed within Newton's 2nd law as the cosmological acceleration factor c^2/R. With this realization, comes the answer to the great cosmological profundity–the ultimate fate of the universe. Without expansion, there is no gravity and without gravity, there is no gravitational collapse. The missing factor in Einstein's opus, the origin of **G**, now revealed as volumetric acceleration of Hubble space divided by Hubble mass. From (J-9), the effective Hubble mass is 1.35×10^{27} kg/meter.

$$G = \frac{Rc^2}{M_U} \tag{J-9}$$

---

[22]How is it, within the limits of experimental error, Hubble mass multiplied by **G** should equal Hubble scale multiplied by 'c^2'? In his search for a scalar-tensor theory of gravity, Robert Dickie would claim the ratio (J-7) revealed the connection between inertial mass and gravitational mass, i.e.,

$$\frac{M_{Gravity}}{M_{inertial}} = \frac{GM_u}{Rc^2} = 1$$

APPENDIX K: ELECTRON ANGULAR MOMENTUM FIELD

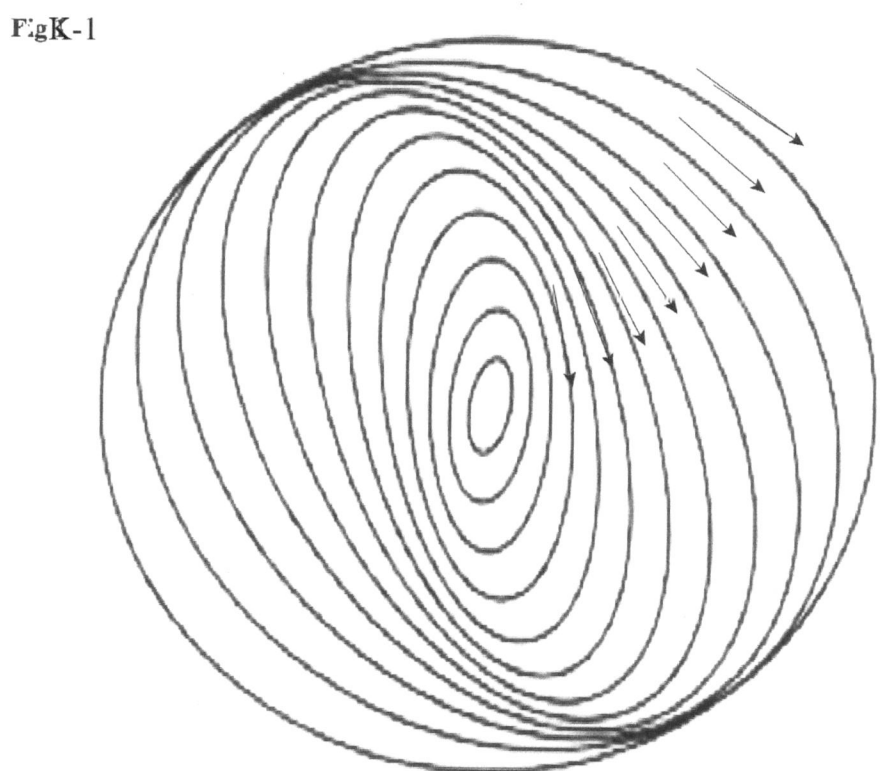

FigK-1

Fig 1 is a 3-D depiction of an electron. It has no substance, it is built from spatial dynamics. In contrast to moving mass vortices [where flow velocity decreases with radius], spatial vorticity velocity is 'c' at all radii [For space to have a physical functionality, $\Delta S/\Delta t$ must equal ('c')]. Bernoulli's equation fulfills the demand for energy conservation along each circulatory streamtube (shown as closed but this is not a requirement). For space, flow is frictionless, consequently all circulations originating within the auspices of the reservoir defined by m_o, will have the same energy. Expansion proceeds as c^2/R, resulting in reactionary centripetal forces normal to all points of all streamtubes equal to $m_o(c^2/r)$ at any point a distance r of the streamtube from the vortical center. The circulation calculated along any closed curve which excludes the center is zero, whereas the circulation calculated along any closed curve that includes the center is $2\pi C$. Ergo, the properties of a spatial vortex are concentrated at a point at the center of rotation, that to which the energy m_o of the circulatory field reflexed.

In his later years, Einstein was asked his thoughts about the short-lived heavy particles that were being discovered (kaons, pions, mesons, etc). Stardard physics viewed this reasearch as an important step in understanding the fundments behind the formula's, they were believed to units of basic matter. Einstein thought differently, and when asked to give an opinion, replied:

"I would just like to know what an electron is."

To Einstin's way of thinking, the answer to the universe lies in its simplicity, which is yet to determined. Understading the pedestran electron (known from the time of the ancient Greeks) was key to comprehending the workings of the universe. Although the 'force' between 'electrons' can be calculated using simple formulas, the equations themselves provide no hint as the structure nor do they shed light upon how force is communicated by charged particles.

In lieu of a discrete physical entity founded upon mass in the form of a sustantive physical property, the electron and its properties are coallased from circuatory space, which at once explains both the electric field and the angular momentum of the electrton as rotational energy. As with earlier work pioneered by Milo Wolff, the rejection of the electron as a discrete material point offers a bountiful harvest. The systematization of subatomic particles as circulatory spatial fields provides a physical interpretation of quantum phenomena heretofore unrealizable.

Mechanical physics acquires a new faculty as circulatory space in that the roles of mass and space are reversed. The contribution of spatial rotation to angular momentum is not $m\omega r^2$ as would be the case if mass were in motion about a rotational center. The electron can be modeled as a vortex [cr_o/r reminiscent of **Fig H-1**)] or as a constant **c** velocity circulation [spread over the circulatory path $2\pi r$ as per **Fig H-2**]. Both formalisms lead to the **1/d** dependence of the angular momentum operative with distance. The challenge is to show that the momentum of the **3-D** circulatory spatial field integrated over the Hubble scale $r_o \rightarrow R$, equals the electron's angular momentum field $h/4\pi$.

The circulation in a plane passing through the eye at radius **r** is cr_o/r. Just as there are two orthogonal planes of spatial rotation contributing additively to the electric force, there are two orthogonal rotational planes that can intersect both the center of rotation and an arbitrary point in space. Spatial angular momentum is therefore the square root of the sum of the squares of two orthogonal circulations multiplied by the central mass m_o, and therefore total moment of momentum LC_T of two intersecting orthogonal circulation planes at any radius **r** is:

$$\begin{aligned}
LC_T &= \sqrt{2}(m_o)\int_{r_o}^{R} \frac{cr_o}{r}(dr) = \sqrt{(2)}cr_o(\ln R - \ln r_o)(m_o) \\
&\approx \sqrt{2}(cr_o)(\ln 10^{26} - \ln 10^{-15})(m_o) \\
&\approx (1.414)(3 \times 10^8)(1.4 \times 10^{-15})(60 - [-35])(9.1 \times 10^{-31}) \\
&\approx 5.1 \times 10^{-35} \text{ meters}^2 \text{kgm} \cdot \text{sec}^{-1} \approx h/4\pi
\end{aligned} \quad (J\text{-}10)$$

The electron electric field **E** and angular momentum field **h/4π**, are alternative manifestations of **3-D** spatial circulation, **q** being but a concatenation of angular momentum. The energy E_p contained in the electric for a point charge **q** is

$$\frac{\varepsilon_o}{2}\int_V \mathbf{E}\cdot\mathbf{E}(dV) = \int_{r=0}^{R}\frac{q^2}{32\pi^2\varepsilon_o r^4} = -\frac{q^2}{8\pi\varepsilon_o}\frac{1}{r}$$

Substituting for q^2 from (H-4):

$$\mathbf{E_e} = -[2\mathbf{m_o}\mathbf{c}^2(\mathbf{r_o})/2\mathbf{r}\,|^{r=R} - 2\mathbf{m_o}\mathbf{c}^2(\mathbf{r_o})/2\mathbf{r}\,|_{r=0}] \qquad (J\text{-}11)$$

Where $4\pi\varepsilon_o = 1/k_e$. (J-11) straightaway revives the problem of particles as points. Clearly, there is no difficulty with the limit **R**, but for **r = zero**, the field energy per (J-11) is infinite. The problem evaporates, however, if the lower limit $r = r_o$ (the charge radius), in which case the energy E_e is:

$$\mathbf{E_e} = \mathbf{m_o c^2} \qquad (J\text{-}12)$$

In the 'inertia-gravity' confluence, negative **G** field energy balances positive inertial energy, i.e., for an electron, the electric field energy (J-13) equals inertial mass m_o. The same quantity of energy is contained in the electric field as the '**g**' field. More interestingly, the circulatory energy m_o may be understood as the refection of the entire circulatory energy to a singular point at the center of rotation. The rotational properties of a perfect fluid are concentrated at its center, circulation calculated along any closed curve that includes the center is **2πC** whereas the circulation calculated along any closed curve that excludes the center, is zero. There is no physical m_o, all is space and the motion thereof.

Circulatory energy m_o per (J-12), defines the angular momentum field **h/4π** (aka the electric field) consistent with the energy in the angular momentum field summed over the Hubble sphere. Negative energy in the gravitational field of m_o equals $(-m_o c^2)$. For the electron, the requirement for zero energy is satisfied by positive mechanical energy in the form of the electro-inertia field $m_o c^2$ and negative gravitational energy $-m_o c^2$ a la spatial divergence. The same thing would be true of the positron, field energy comprises a different order of rotation as the distinction, but field energy is always negative relative to positive mc^2 energy. There are no alternative constructs for electrons and positrons. All electrons are alike because the energy relationships do not permit a variance. Only those values of circulation fit the universe at a particular era.

FIG K-2

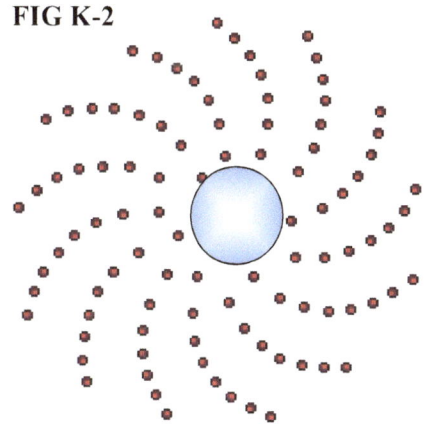

Fig K-2: A global slice taken through an electron depicting spatial angular momentum influx **h/4π** (red) entering the charge sphere (black) representing the reflexed energy of the vortical spatial influx $m_o c^2$. Electrons cohere as contra angular momentums, one related to charge radius $(m_o c r_o)$, the other $(h/4\pi)$, being the global extension of the electron spatial angular momentum field (aka electric field) obtained by integrating over the Hubble volume (J-10).

Appendix P: Particles as Circulations Spatial Circulation

"The most shocking and disturbing thing about quantum mechanics is that if you take the angular momentum along any particular axis, you will find it is an integer or half integer multiple of $h/2\pi$."

Richard Feynman

How can a particle simultaneous manifest as angular momentum in all directions? Richard Feynman (the most gifted of educators in reducing mathematical complexities to simple analogies) would eventually lament: *"Understanding these matters comes slowly, if at all."*

As was the case with the demystification of gravity, it is the demonstrative subliminalities of inertially fortified spatial dynamics that will provide the agent of action for the electrical and nuclear force. The void again becomes both mediator and player, here now in the micro-arena of the subatomic saga. The distinction between quantum angular momentum (as circulatory space) and classical angular momentum (as rotating mass), alas resolves Feynman's enigma. Spatial circulations are 3-D. The virtual mass-energy of an angular momentum system resides at the circulatory center. As is the case with gravity, Newton's 2nd law is operatively symmetric, there is no difference between force created by acceleration of space wrt to mass and acceleration of mass wrt space.

The electric field shown in **Appendix H** is emulate-able as an ideal spatial vortex spiraling inwardly to reach 'c' velocity as **r** reaches r_o, or as distributed space having 'c' velocity at all radii. Shown in **Fig 11-A** and **Fig 11-B**, the constitution of an electron [blue] or positron [red] reduces to a reflexed central quantum mass illusory placeholder m_o, a spatially distended 3D velocity field '**c**,' and a dimension interface of radius r_o defined by the reflexed energy m_o of the circulatory field. The spatial field at any point '**P**' at distance '**d**' in any direction is ⊥ to the radial line drawn from the central core to the point, but its direction in the surface plane ⊥ to the radial will depend upon the polarity of the assembly that defines the particle as either an electron or positron. There need not be a physical inwardly spiraling spatial flux, although the mathematical vortex leads to the same expression. Taking a slice through the **X-Y** plane when the electron and positron are spaced apart a distance '**x**' as shown in **Fig 11-A**, their circulatory spatial fields act upon one another to create the attractive forces developed in **Appendix H**. If the positron and electron could be brought together in a close quantum linking configuration as shown in **Fig 11-B**, the oppositely circulating fluxes cancel on average. If they could be superimposed without destroying one another, the fields totally cancel.

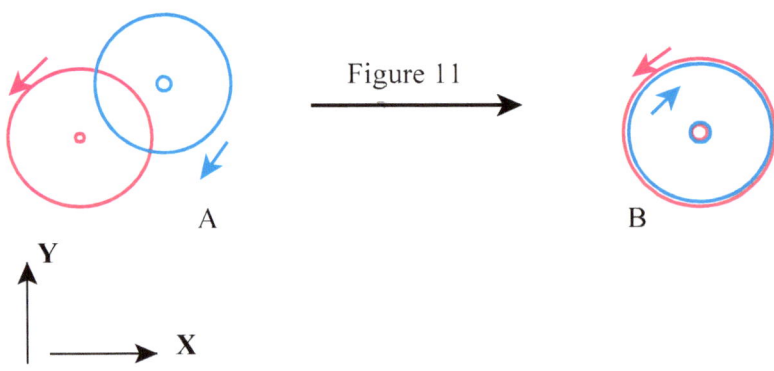

Figure 11

While an electron and positron cannot exist as a combined entity, other charged particles (such as the proton and electron, can exist as a unit, for example as a hydrogen atom. Because there two circulatory fields are not precisely concentric, the vortical centers rotate about a common center. Consequently, particles formed from a combination of single spin particles exhibit 3-D residual standing waves as shown in **Fig 11-C** ---------------------------------->

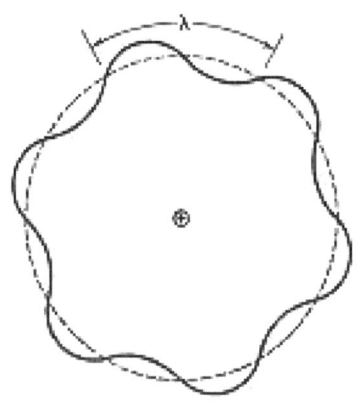

The resulting wave will not have a polarity but it will have an oscillatory amplitude in the range of the center to center spacing between hub mass m_o. Two oppositely circulating 3-D polarity defining spatial fields [hereinafter the angular momentum electric charge] of each particle, cancel on average, but as above explained, because the electron and proton rotate about a common center of mass, they create a stationary wave as shown in **Fig 11-C**. The wave nature of chargeless particles discovered by de Broglie, can be attributed to the slight displacement between the cancelling circulatory electric fields.

When opposite charges are combined as a single particle, matter appears in form as particles without polarity. Of interest here, is the binding force created when vortices link at close range to create subatomic bonds   The electron exists in two guises, 1) a free unbound state having mass m_o, and radius r_o, and 2) a bound state as a (mu minus) having mass m_u (approximately 207 times that of the unbound state) with a correspondingly smaller classical radius. The strength of the acceleration field in the free state has a maximum value c^2/r_o at the Hub. Circulatory space is responsible for all subatomic binding actions as well as the long range the electric fields[23]

Gravitational and electrical forces have measurable coefficients **G** and **q**, so new theories of gravity and charge can be quickly discredited if they do not produce the correct force. The strong force, by contrast, is not known with precision. It depends from complex geometric(s) involving rotational and positional uncertainties.

The present excursion into the subatomic world will be limited to an adaption of circulatory flux to compact spaces. This reinvigoration of the angular momentum flux on a Lilliputian scale leads to a simplistic overview of subatomic binding within the spirit first proclaimed by William of Occam.

---

[23]In 1933, Hideki Yukawa hypothesized nuclear binding potential could be related in form to the Coulomb force if multiplied by an inverse exponential time delay function conflated to reflect a mass factor approx 200(m_o):

"It seems natural to modify the theory of Heisenberg and Fermi in the following way. The transition of a heavy particle from neutron state is not always accompanied by the emission of light particles. The transition is sometimes taken up by another heavy particle."

The μ meson (207m_o) now called muon, was initially thought to be Yukawa's particle, but as later shown, it did not fit the *standard theory* of forces that evolved around the idea of spin one virtual particles traveling between nucleons. Somewhat heavier mesons called pions were subsequently discovered and adapted to the standard theory, leading to a Nobel prize for Yukawa. While not a part of the *standard model,* muons are key to circulatory coupling theory

Appendix Q - Matter From Rotational Space.

Photons have angular momentum **h/2(π)** but no measurable mass. Electrons have mass m_o woefully insufficient to explain the electron's intrinsic angular momentum **h/4(π)**. The photon angular momentum is indicative of circular rotation whereas electron angular momentum bespeaks of a 3-D spherical distribution. There is a unique relationship between the electron mass energy m_o and the mass energy of a gamma ray photon of wavelength $\lambda = h/m_o c$. The photon energy $E = hf = m_o c^2$ Photon momentum-energy depends from '**c**' (photon velocity wrt space). Likewise, electron mass-energy m_o also derives from **c** (the circulatory velocity of space wrt mass).

The energy reflexed from the spatial circulatory vortex to the charge sphere is that which corresponds to the $m_o c^2$ energy required to move infinitesimal elements of charge from infinity to a spherical surface. The essence of the electron, indeed, the root cause of what is commonly believed to be the fundamental unit nature called "charge," is but a confederation of local and global dynamic factors (specifically charge sphere radius r_o the vortical energy m_o in the circulatory spatial field, and the spatial expansion factor c^2/R).[24]

Charge is not a fundamental entity, but rather a naturally occurring extemporized relational charade. The metaphorical process of charging an imaginary sphere to equal the empirically determined mass-energy m_o, provides the analogy. Once m_o and r_o are known, the concept of charge reduces to a 3-D spatial angular momentum vortex $h/4\pi$ having a total energy mo energy per (J-12). In words, the integral of the electron energy from r_o to **R** (the volume of the Hubble sphere) is m_o. The true source of m_o is thus dynamic (the c^2 factor being a result of the spatial expansion factor c^2/R). The notion of "**q**" as a conceptual quantity of charge, having well served a valuable purpose for more than two centuries, can now be understood as the resistance field created by the expansion of angular momentum. At the r_o interface, inwardly directed acceleration is c^2/r_o. Consequently the following relationships apply:

$$-P = \frac{c^2}{r_o} \times \frac{m_o}{4p\,r_o^2} = \frac{m_o c^2}{4p\,r_o^3} = \frac{m_o c^2}{3 \times V_o} \qquad (Q\text{-}1)$$

$$E = -3PV_o = m_o c^2$$

$$-P = \rho_U c^2 / 3$$

All electrons are alike because reflected $m_o c^2$ energy derived from the expansion of the vortical angularmomentum field **h/4π** must be satisfy (Q-1).[25]

---

[24]The energy corresponding to $m_o c^2$ is that which determines r_o. The charge sphere comports to what is required to build the electron. There is only one value of r_o that comports with m_o. Specifically, $(1/2)e^2/r_o = m_o c^2$. If the effective radius were more or less than r_o it could not simultaneously satisfy the $-m_o c^2$ gravity field and the $+m_o c^2$ of the electro-inertial field.

[25]The vortical eye correspond to the "charge radius " of the particle., wherefore the classical **v x r = k** vortex reaches '**c**' **x 0**. We are now able to define some particles in terms of electron like circulatory fields.

Appendix R -- The Sommerfeld Fine Structure Coupling Constant "α"

The fine structure constant, *Alpha*, has been a puzzle since its discovery. Its numerical value is approx (7.297×10^{-3}):

$$\alpha = [q_e^2/4(\pi)(\varepsilon_o)hc] \tag{R-1}$$

but usually expressed as $\alpha^{-1} = 137$. When first introduced by Sommerfeld as spectral line splitting, *Alpha* was satisfactorily rationalized as the ratio of the speed of light to the electron speed in the first Bohr orbit. The question soon arose as to why these factors reduce to the ratio of two velocities? Being dimensionless, it makes sense numerator and denominator represent factors having the same units, but what is special about the velocity of the electron in the first Bohr orbit?[26] Traditionally regarded as having an electrical pedigree, and therefore dependent upon electromagnetic parameters, can now be understood as the origin of the enigma. The most common and constants in quantum electrodynamics ('**c**' and '**h**') describe mechanical properties (velocity and angular momentum).

Alpha subsequently evolved into the mystery that still remains, its bearing upon the strength of the electric field q^2.[27] This, however, is not obvious from its dimensionality since electric and magnetic forces fall off inversely with distance To consider *Alpha* as a dimensionless force, one needs to consider the proportionality factor $e^2/4\pi\varepsilon_o$ and substitute **hc** for the inverse dependence of force upon distance. This fits the *Alpha* factor into the same format as those used to describe the strong force coupling constant (approximately 1) the electric coupling constant (1/137) and gravity (10^{-41}). The problem is, that $e^2/4\pi\varepsilon_o c$ is not recognized as angular momentum.

From the Bohr model of the hydrogen atom and (H-7), the velocity in the first orbit has a value:

$$v = k_e(q_e^2)/(h/2\pi) = 2m_o c^2 r_o/(h/2\pi) = c[m_o c r_o]/(h/4\pi) \tag{R-2}$$

which is the ratio of two angular momentums multiplied by **c**. The numerator is determined by the counter momentum field of the spherical Hub, the denominator defines the global angular momentum field of circulatory space (i.e., the electric field), and (H-7) is the Rosetta Stone between electrical and mechanical physics. The ultimate question of charge as a fundamental property of nature has been answered in the negative. Electrons are the natural constructs of dynamic fundament(s). Electron force is the result of angular momentum opposition to acceleration. The circulatory spatial field summed over the Hubble volume —> $h/4\pi$. Since the factor $k_e q_e^2$ is the same as $q_e^2/4(\pi)(\varepsilon_o)$, then:

$$\alpha = \frac{q_e^2}{4\pi\varepsilon_o (hc)} = \frac{v}{c} = \frac{k_e q_e^2}{hc/2\pi} = \frac{2c(m_o c r_o)}{ch/2\pi} = \frac{m_o c r_o}{h/4\pi} \approx \frac{1}{137} \tag{R-3}$$

Expressed in terms of its mechanical pedigree, Alpha reduces to the ratio of two angular momentums.

---

[26] Actuality, *Alpha* can be represented in several ways depending upon which constants are chosen.

[27] Wolfgang Pauli once quipped "When I die, my first question to the devil will be:

"What is the meaning of the fine structure constant?"

Appendix S – Spatial Influx Particle Formation - Negative Pressure Sink

The energy in the **g** field of an electron is by theory and measurment $m_o = 9.1 \times 10^{-31}$ kg. There being no other measurable inertial, mo must also represent the energy in the electric field energy. In other words, there is only one mo energy factor and it derives from rotational space. Indeed there is no explanation for the electron mass other than that it represents the dynamic energy of spatial rotation. It is by this ruse, nature hoodwinks its investigators. Gravity fields are always balanced by, and the result of, positive mc^2 energy sources, at least that is what has been assumed. But there is at least one alternative interpretation of nature, specifically that it is the field that is primary. The inertial of mass is perhaps due to the difficulty in accelerating a field coextensive with the Hubble sphere. While the properties of the vortex are concentrated and experienced as existent at its rotational center, the reality may be quite different. The charge radius r_o and its premised co-rotating central mass m_o may both be illusory. If the source of the electron's inertia is in the field, then the circulatory field is primary - the electron is not a point, but rather a large volume of expanding circulatory space.

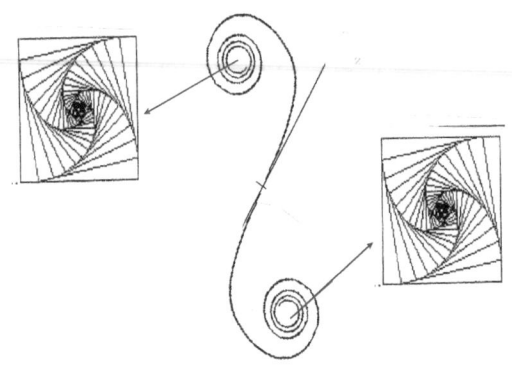

Electrons and Positrons are always created in pairs. Net angular momentum is therefore zero from a cosmological accounting perspective. Both the electron and positron are considered negative pressure sinks (as are all manner of masses and energies that create external **g** fields). Thus cosmic angular momentums is balanced by the simultaneous creation of electron-positron pairs as shown

(**Fig S-1**), ---->

The balance between positive $m_o c^2$ energy and negative circulatory field energy $m_o c^2$ must be accounted for within the structures themselves. The mass mo which is the bases for the gravity feild of the electron is one-in-the-same as the energy in the energy in the global angular momentum field $h/4\pi$ and the angular momentum field $m_o c r_o$ of illusory. The reflection of the global angular momentum field $h/4\pi$ to the rotational center is a root cause of the alpha enigma. That the electron's positive $m_o c^2$ energy be balanced by negative $m_o c^2$ field energy, the electron's **g** field mass exists in the form of rotational space. There is only one radius r_o where the mass energy m_o of the global spatial circulation will equal the mass energy m_o of the charge radius r_o. The sum of the negative circulatory energies contained in the **E** field (aka the angular momentum field), must equal m_o per (J-12), and this must at once correspond to the energy m_o contained by the surface area ($4\pi r_o^2$) of the charge sphere. Note, however, the illusory mass m_o, need not physically rotate in order to produce counter angular momentum. Counter angular momentum $m_o c r_o$ is created by the static state of m_o wrt the directional influx of circulatory space at the r_o interface as shown in **Fig S-2** ....>

The pressure flux energy [(S-2) below] contained in the electrons spatial angular momentum field (J-10) accords with the integration of the electric field energy per (J-12) summed over the Hubble volume from r_o to R. Taking m_oc^2 field energy per (J-12) as existing in form as spatial circulation akin to **Fig K-1** (circulatory flow energy of the angular momentum electric field), the electron reveals as two concentric counter circulatory angular momentums $h/4\pi$ and m_ocr_o. That the mass m_o created by the global extent of the spatial circulatory field is reflexively transformed to the concentricity center of the charge sphere, it functions both as virtual mass counter momentum m_ocr_o and positive mc^2 to counterbalance the electron's negative **g** field energy (as required for all residents of a null universe).

The uniqueness of the m_o - r_o conjugation is critical to the existence of long range electric fields. Electrons and positrons occupy a precise niche. The negative and positive energies of each particle is internally balanced, but angular momentum nullification depends upon particle pair creation.[28] This has a profound effect upon the notion of electrons and positrons as fundamental descriptors of electric charge. Were it not for the intrinsic angular momentum imbalance between circulatory space $h/4\pi$ and virtual counter angular momentum m_ocr_o, there would be no fine structure constant "alpha." Moreover, if the entire angular momentum source field $h/4\pi$ where taken-up by a local counter spatial circulation, as is the case with most other particles, there is noting left of the long rage electric field. Angular momentum spatial circulations are the source of electric fields and the binding force. In the case of electrons and positrons, most of their angular momentum spatial circulation fields is dedicated to the the long range electric field. In the case of the strong nuclear binding force, however, the greater part of a particles circulatory field is inter-linked with the circulatory spatial fields of other particles to form retaining bonds between nucleons.

To complete the electron model as a source of force, the expansion field must act upon the angular momentum field in a manner that creates reactionary force. In analogizing the interface between the charge sphere and circulatory space to that of the interface between non-expanding matter and isotropic spatial expansion, spatial influx again provides the momentum flow (already a part of the inflow similitude in the form of the vortical spatial flow.

To repeat, there is only one dimension distance [r_o] where the negative mass-energy [m_o] of the electron's spatial angular momentum field can equal the positive [m_oc^2] energy of charge sphere. Integration over the volume of the Ee field establishes mo and consequently the cosmological pressure at the surface of the charge sphere. The expansion rate c^2/R creates the pressure **P** at all locations within the circulatory spatial flux. As previously, field pressure is simple the product of the spatial acceleration field multiplied by the area density of the circulatory field $m_o/4\pi R^2$:

$$P = \frac{c^2}{R} \times \frac{m_o}{4\pi R^2} \tag{S-1}$$

Hence
$$E_e = 3PV = m_oc^2 \tag{S-2}$$

---

[28] If the circulatory spatial field were momentarily allowed to create greater energy, the charge radius would expand, counter angular momentum would increase, resulting in a decrease of spatial circulatory momentum. The electron is a self stabilizing negative feedback condition which maintains all electron existence within the boundaries of the niche. Consequently, all electrons are alike. Perhaps it is time to reconsider John wheeler's cherished view: "*all matter is made of electrons.* Or those of J.D, Ross: "*Mater is a mirage, all is space and the distortion thereof.*"

Cosmological expansion creates negative pressure stress fields that can now be identified with particle mass. Expansion is not only the source of gravity and the energization of mass, but it also functions as the source of charge. The common electron is key to understanding matter. All particles that can exist (even momentarily), have spatial fields by which they communicate with the universe. In the presence of other particles, those fields may cancel, reinforce, interlink, or interact to repel or attract. On the subatomic scale, they are the bonding mechanism, on the global; scale they are the Electro-Angular Momentum field

Gravity is the pseudo force pressure created by the action of spatial expansion acting upon expansion resistant matter. Charge is the result of expansion acting upon expansion resistant angular momentum.

Appendix T -- A Short Essay on Spacetime Curvature

A recurring question posed by the prophesied curvature of spacetime, is that of how Einstein's relativistic transforms mediate the path of Newton's free-falling apple. That the apple follows a geodesic trajectory coincident with earth's center, will be recognized as a facet of the general issue as to how intangible spatio-temporal distortions supplant Newton's **g** force?

Except for bodies of extremely high mass, the degree of spatial distortion is *de minimis*. What then did Einstein have in mind when he substituted spacetime distortion for Newtonian gravity. For most celestial objects (ordinary stars, planets and moons), the effect of mass **M** upon spacetime is solely temporal. That clocks run slow in a gravity potential is well documented, but is the path of a falling apple determined by the rate at which clocks log time, or directly by the energy gradient created by **M**. If clock slowing is collateral rather than primary, the energy state of space becomes paramount. While spatial curvature could be resurrected as a surrogate under the auspicious of variegate energy density, the foundation basis for gravity as spacetime curvature, loses credibility. The free-fall options are then restricted to '**g**' force physics *a la* Newton or as potential energy gradient.

In Newton's world, earth's mass creates an acceleration force deemed to act directly upon other masses. In Einstein's universe, **M** creates a potential energy well. That **M** also determines the time dilation factor in the relativistic field equation, the spatial state can be thought of as a time dilation gradient wherefor the free fall path can be considered either as a consequence of time dilation or as spatial rate of change of energy. In either case, the need for a **g** force is abrogated

The potential energy field outside a non-rotating sphere of mass **M** slows "time" in accord with the predictions of the General Theory by a factor:

$$\Delta t_o = \Delta t_i \sqrt{1 - \left[\frac{2MG}{yc^2}\right]} \quad (1)$$

where Δt_o is time passed by a clock in the gravity field of **M** at distance **y** from the center of **M** and Δt_i is the time passed by a clock at an infinite distance from **M**. The factor $[2GM/y]^{1/2}$ is the escape velocity **v***. Clocks far removed from the influence of mass run faster. A clock on earth's surface (distance **r**) runs at a slower rate than a clock at distance **y**.

The square of the escape velocity $(v^*)^2$ corresponds to the kinetic energy required to escape the influence of **M**. Ergo, $(v^*)^2$ and consequently potential energy, diminish with distance from the earth's center.

$$(v^*)^2 = 2\frac{MG}{y} \quad (2)$$

And the energy gradient is:
$$a = \left[\frac{d}{dy}\right]\left[\frac{2MG}{y}\right] = \frac{2MG}{-2y^2} = -\frac{MG}{y} \times \frac{1}{y} = -\frac{(v^*)^2}{y} \quad (3)$$

Classical physics thus offers two formalisms for Force, both produce the same acceleration MG/y^2
1) Rate of change of momentum, **F = dp/dt**, and 2) Rate of change of energy with space **F = dE/dS**

Appendix U – Speculations on Unification of Electro-Mechanical Impedance

From Friedmann's 2nd equation:

$$-P = \frac{\rho_U c^2}{3} \tag{1}$$

And since

$$P = \frac{\sigma_U c^2}{R} \tag{2}$$

Then

$$\frac{\sigma_U c^2}{R} = \frac{\rho_U c^2}{3} \tag{3}$$

Hence

$$\sigma_U = \frac{\rho_U R}{3} \tag{4}$$

From Maxwell's equation

$$c^2 = \frac{1}{\mu_0 \varepsilon_0} \tag{5}$$

From (2) and (5)

$$\sigma_U = \frac{PR}{\mu_0 \varepsilon_0} \tag{6}$$

Appendix V. Musings On The Strong Coupling Force

> *"If we are not content with the dull accumulation of experimental facts, if we make any deductions or generalizations, if we seek for any theory to guide us, some degree of speculation cannot be avoided."*
>
> Arthur Eddington

Gravitational and electrical forces have measurable coefficients **G** and **q**, so new theories of gravity and charge can be quickly discredited if they do not produce the correct force. The strong force, by contrast, is not known with precision. It depends from complex geometric(s) involving rotational and positional uncertainties.

The present excursion into the subatomic world will be limited to an adaption of circulatory flux to compact spaces. This reinvigoration of the angular momentum flux on a Lilliputian scale leads to a simplistic overview of subatomic binding within the spirit first proclaimed by William of Occam. For a self repulsive field, larger mass comports with smaller size. From theory supported by experimental findings, the muon charge radius is indeed less than r_o, but as developed herein, factors other than mass come to play in determining natures short range operative. So while classical reasoning points to a theoretical muon size, the effective size of the muon as an instrumentality of comparison will depend upon several complexities.[29] To achieve strong binding forces from circulatory interaction, flux energy must be compacted into volumes commensurate with the cross sectional area of the muon itself. This can take form either as a density endowed spatial envelop or rotational momentum of the particle as circulatory flux.

In deriving the coupling constant '*alpha*,'' (Appendix N), two angular momentums were identified. (local pseudo angular momentum $m_o c r_o$) and spatial circulatory angular momentum $h/4\pi$. Pseudo angular momentum is defined by the electron's effective charge radius r_o and the reflected mass-energy m_o attributable to the spatial angular momentum factor $h/4\pi$. Muonic mass-energy is greater than m_o by a factor of 207. In the free state, the muonic local flux equals $h/4\pi$ the muonic coupling factor—> unity. When circulatory flux binds to another particle with the full momentum of the $h/4\pi$ circulatory field, the strong force coupling constant is 137 times greater than the electron coupling constant α. While the electron's coupling factor is counter angular momentum created by the electrons spatial momentum field, the muon coupling factor diverts spatial circulatory field to binding force that sustains a preliminary the above estimate of nuclear binding strength. Particles are postulated to experience the angular momentum fields of other particles when the lines of motion are captured within the effective rotational radius of the other. If the muon is taken to be a fundamental particle (i.e, a shell having no internal structure), its $h/4\pi$ angular momentum is accommodated by an effective radial rotational length r_e:

$$r_e = (5.3 \times 10^{-35})/(207)m_o[c] = 0.93 \text{ fermi}. \tag{2.2}$$

---

[29] Much effort has been directed to endorsing certain actions, processes and particles, not because they are proven by the experiments, but because they are not excluded by the results.

The assumption of local counter angular momentum based upon the mass factor $m_\mu = 207m_o$ leads to a momentum transference rate in the range of the estimated strength of the strong force. For present purposes it is only necessary to consider the muonic coupling mechanism as short range, the effective momentum of momentum for purposes of binding adjacent particles being confined to an area defined by the muon's disposition within a subatomic composition as more fully developed below. The rate of change of momentum will be moderated by the momentum flow (operative density multiplied by **c**). Muonic flux flow determines the force.[30]

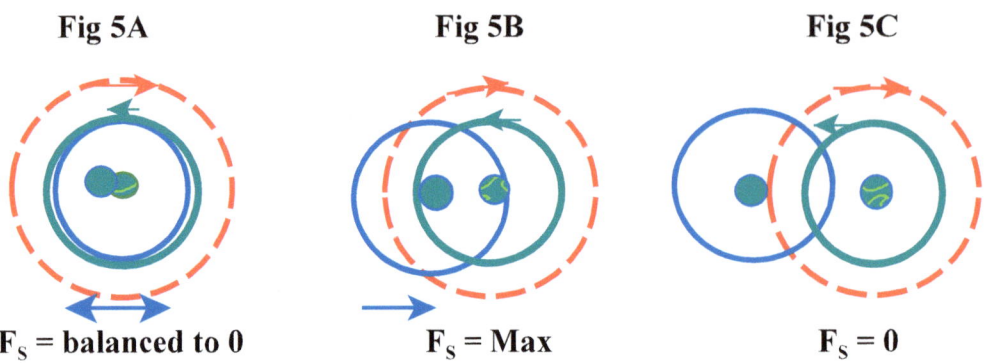

Fig [5A] depicts two particles Green and Blue each shown with two circulatory spatial planes. One circulatory plane of each particle is in the plane of the page (the green and blue circles). The other circulatory plane is normal to the page (shown as green and blue flux bundles centered within the green and blue circles). The large red circle is where the electric field of the green particle begins. When flux centers are coincident (blue bundle coincident with green bundle), the binding force is minimal. In Fig [5B] the blue particle is shifted to the left. As it is moved the green centering field flux increases. The further the displacement, the greater the restoring force of the green flux. In [5C] the blue particle is unbound (beyond the counter rotating centering flux of the green particle and the linked bond is severed.

Fig 6 binding forces are confined to small areas and consequently the coupling flux is highly concentrated. The retentive force that holds nucleons together and indeed, the elements that make up nucleons, are testament to the high density circulatory interaction sustained by the angular momentum flux -->

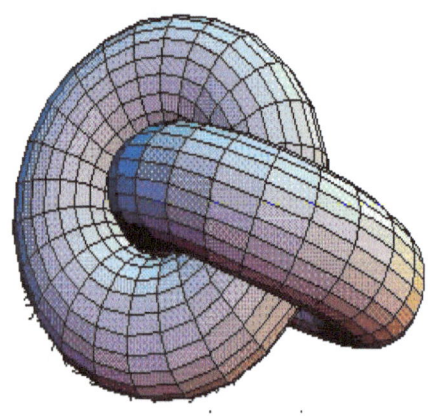

---

[30]High density spatial environments mean high pressure and pressure augers compression. If pressure transfigures electrons to muons then stress emancipated muon's relax to electrons.

Quantum theories of force are grounded upon all or nothing impact exchange. Boson(s) are never seen, and there is no accounting for their numbers. Annihilation is without regard for conservation. Real particles, by contrast, are endowed with measurable properties. Charge, baryon number (protons + neutrons) and angular momentum is conserved no matter how many new particles are created in a process. When the mass **m** of a particle is sufficiently large, the alpha factor is unity and counter rotational velocity **c** corresponds to angular momentum **mcr = h/4π**. For the muon, $\mathbf{m_m}$ (**207$\mathbf{m_o}$**). This corresponds to a rotational radius $\mathbf{r_e}$ = **0.93 fermi** per (2.2). The question arises then, for a unity coupling constant, what is the force. The maximum electrical acceleration is $\mathbf{c^2/r_o}$ and action of this acceleration upon he electron mass is $\mathbf{m_o c^2/r_o}$, so by like reasoning the action of the muon acceleration $\mathbf{c^2/r_e}$ upon the mass of another muon would be 320 times greater than electric force.

$$\mathbf{F_e = m_m c^2/r_e} \approx 320 \qquad (2.3)$$

Mechanical physics acquires a new faculty as rotating space, the role of mass and space is reversed. The contribution of spatial rotation to angular momentum is not $\mathbf{m\omega r^2}$ as would be the case if mass were in motion about a rotational center, rather it diminishes as $\mathbf{cr_o/r}$ reminiscent of a free vortex velocity. Thus, while there is a constant velocity **c** associated with circulatory spatial flow at all radii, the contribution at radius **r** is distributed over the spatial length **2πr**, and since this length increases with radius, the effect of const '**c**' circulation is inversely dependent upon the radius.

As developed in *Appendix J* in connection with the electron, the ultimate challenge for any theory is "What does it predict?" In the present context of that objective, it will be to show that the spatial circulatory electric field corresponds to the angular momentum **h/4π** when integrated over the volume of the Hubble sphere. The circulation in a plane passing through the eye at radius **r** is $\mathbf{cr_o/r}$. As there are two orthogonal planes of spatial rotation contributing additively to the electric force, there are two orthogonal rotational planes that can intersect both the center of rotation and an arbitrary point in space. Spatial angular momentum therefore equals the square root of the sum of the squares of two orthogonal circulations multiplied by $\mathbf{m_o}$, that is:

$$\mathbf{LC_T} = \sqrt{2}(\mathbf{m_o}) \int_{r_o}^{R} \frac{\mathbf{cr_o}}{\mathbf{r}}(dr) = \sqrt{(2)} \mathbf{cr_o} (\ln R - \ln r_o)(\mathbf{m_o})$$

$$\approx \sqrt{2}(\mathbf{cr_o})(\ln 10^{26} - \ln 10^{-15})(\mathbf{m_o})$$

$$\approx (1.414)(3 \times 10^8)(1.4 \times 10^{-15})(60 - [-35])(9.1 \times 10^{-31})$$

$$\approx \mathbf{5.1 \times 10^{-35} \, meters^2 \, kgm \cdot sec^{-1}} \approx \frac{h}{4\pi}$$

Appendix Z: Intergalactic Communication -> From Here to Infinity

Fig 1 shows a square laminate ⊥ to the **Z** axis having area **R²** commensurate with the area of a great circle through the Hubble sphere. **Fig 2** adds a plane parallel to the **Y-Z** axis, and a plane parallel to the **X-Z** axis. In the real universe all space is filled with cubical blocks to form a solid cube as which can be viewed from either the X, Y or Z axis as plenum of parallel planes orthogonal thereto. However, because the **g** field of large planes is perpendicular (⊥) to the plane, the three dimensional inertia of the universe can be represented by the three orthogonal planes shown in **Fig 2**. In makes no difference where the planes are located along the axis, e.g., the **g** field of the **x-y** plane is parallel to the **Z** axis and has the same intensify at any distance. Dido for the **x-z** plane and the **y-z** plane, there is no diminution of gravitational intensity with distance.

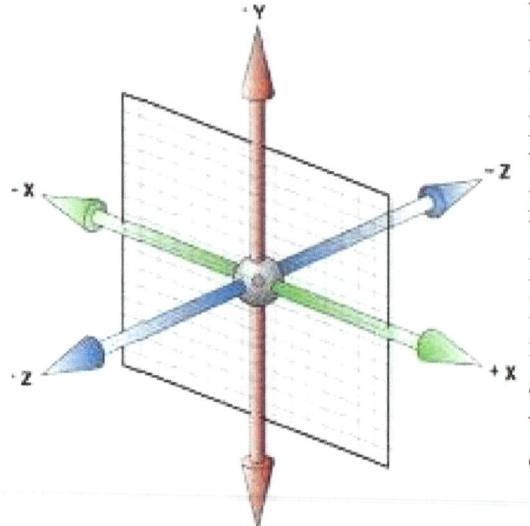

Fig 2 represent the inertia of the universe - the inertial reaction of any accelerated body is retro reflected from that plane which is orthogonal to the acceleration. The actual orientation and location of the coordinate system with respect to the universe and the acceleration of a body is immaterial. The universe always manifests as a ubiquitous retro-directive infinite plane area density σ_U to each and every element of an accelerated inertial body. Rationalization thereof, follows as a necessary consequence of the non-compressibility of negative pressure space which communicates the sum total of all cosmic mass in form as a plenum of infinite-area planes in an infinite universe (or a plenum of large planes in a finite Hubble sized universe). The fact that σ_U has a finite value, would ostensibly rule out the former interpretation at least in terms of its total matter content. It is possible of course, that mass and the expansion of space upon which mass depends, constitute only a finite volume within an infinite volume of empty non-expanding space. Whatever the case, Hubble parameters are in harmony with locality.

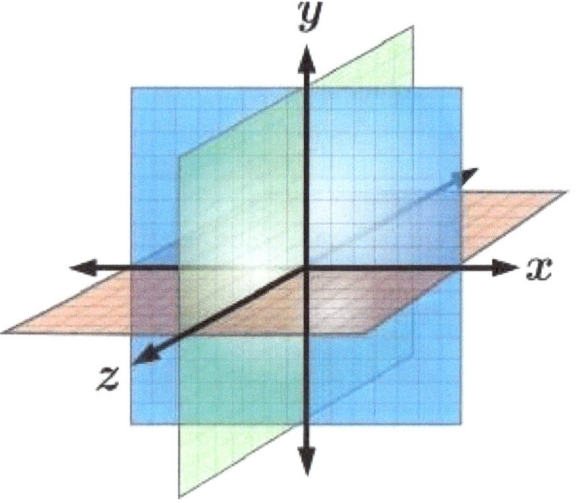

Per **Fig 2**, each of the three coordinate planes separately represents a one third share of the Hubble mass. Operatively, only that plane normal to the direction of an accelerated mass is involved in creating inertial reaction. The other two planes do not participate (inertial reaction is confined to instantaneous anti-parallel non-divergent counter action. The physiology of space as approximately homogeneous volume density hides its true office as retro-directive area-density impedance.

That each of the three planes represents all planes parallel therewith (into which the Hubble universe can be sliced). To determine area density σ_U from the volumetric density ρ_U, express both in terms of the Hubble mass M_U:

$$M_U = \rho_U(L^3), \text{ and } M_U = \sigma_U(3L^2)$$

Hence:
$$\sigma_U = \frac{\rho_U(L^3)}{3(L^2)} = \frac{\rho_U L}{3}$$

Understanding the universe as an area density explains Newton's 2nd law, and consequently his law of gravity as inertial reactance of mass imposed upon the isotropic spatial expansion field. Space is indeed conditioned, as Einstein opined in 1916, but not in the way he surmised.[31] There were no known physics principles for explaining mass induced spacetime curvature in 1916, nor do they exist today. What does exist in form both then and now, as the global **G** field, does not conform to Einstein's expression for static curvature of static space, but rather spatial divergence (cosmological expansion) implied by the *ad hoc* cosmological constant Λ.

From Gauss's Law, the **g** field of an infinite plane is he integral over a flat surface **S** of area "A" is:

$$\oint \mathbf{g} \cdot \mathbf{n} \, da = -4\pi GM = g(2A)$$

Fig 3: Force lines as belabored above, are perpendicular to a surface **S** defined by the cross sectional area of a cylindrical hole imaged as passing through a continuous plenum of parallel planes representing one dimension of Hubble expansion (that parallel to the cylindrical axis). Total area is **2A** (both ends of the cylinder), and total acceleration flux is twice the radial influx ($4\pi G$) at one end. Area "A" equals the two ends of the cylinder. The mass **M** is that enclosed by the cylinder (area **A** multiplied by the area density σ_U. Hence:

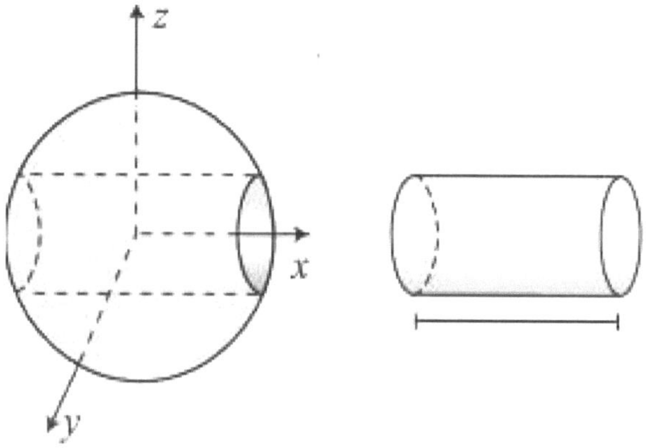

$$g(2A) = [-4(\pi)G(2A)]$$

Therefore:
$$g = -4\pi G \sigma_U$$

That the **g** field at all places within the cylinder is constant and equal to $4\pi G\sigma_U$, the effective inertial impedance at all points therein is

$$\sigma_U = g/4\pi G = g[4(\pi)R\sigma_U]/4(\pi)c^2$$

Whence:
$$g = c^2/R$$

---

[31] There being no known physics for orchestrating gravitational effects from static space, Einstein seized upon the idea that mass could alter spacetime in such a way as to explain the curved paths of moving bodies in the vicinity of large mass. It is somewhat surprising that Einstein would adopt a static theory of the universe that needed the protection of a hypothesized dynamic factor [Λ] to prevent gravitational collapse. As it appeared in the final draft of his General Theory, the cosmological constant Λ is multiplied by **R/3**, hence Λ would need to have the value **3H^2** to balance gravity on the global scale.

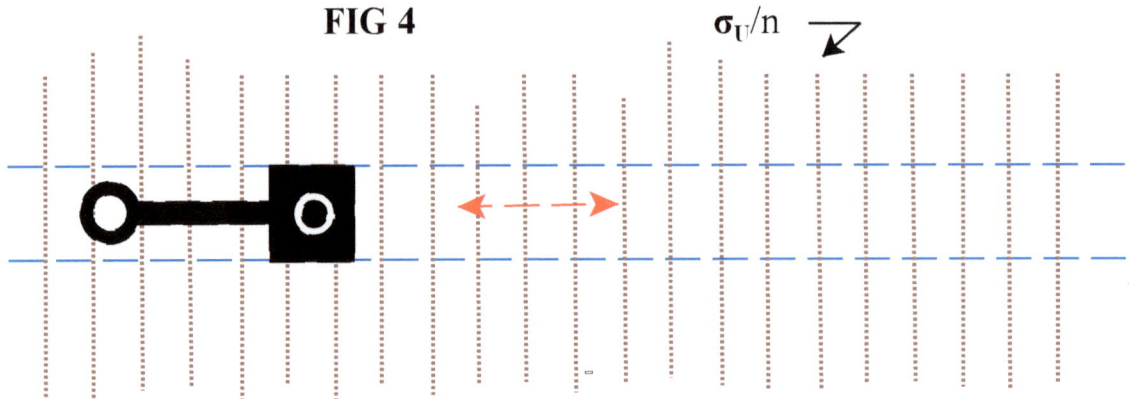

Fig 4: An oscillating piston (red arrows) moves back an forth within the factitious cylindrical confines of the path defined by thereby, There is of course no real cavern or hole within which the piston moves, the universe being substantially empty, the planes themselves are largely devoid of material bodies. [The absence hole boring machinery on the scale of the universe is not an obstacle to our thought experiment]. The back and forth motion of the piston is opposed by the plenum of planes into which the universe has been imaginarily sliced (dotted vertical lines). The divisions are metaphorical, for the tutorial purpose of illustrating the cosmodynamic reaction to the oscillatory accelerations of the piston. Each plane contributes an incremental laminate of mass to the cumulative virtual area-density σ_U. The piston always feels the area density as constant opposition to acceleration. Whether the constant cosmological area is an existing local condition or instantly transmitted non-compressible pressure wave, the cumulative action is the same.

That the universe acts to oppose acceleration of the piston, the piston acts upon the universe. Each plane feels the acceleration and accordingly, an oscillating body will be felt by all planes contributing to the cumulative **g** field, That each plane contributes only σ_U/n units of mass, it will experience only σ_U/n fraction of the force amplitude (**F = Ma**). Because inertial reactance is instantaneous, the accelerating motion of the piston is communicated instantaneously to all mass which contributes to the **g** field within the cylinder.

Detecting modulation superimposed upon the cosmological 'g' field is beyond the present state of technological imagination. The global 'g' field is the product of Hubble reactance multiplied by Hubble expansion. That Inertial reaction is instant, it is of academic interest.

Recapitulation

In the Cosmodynamaic construct, spatial expansion creates global divergence. The difference in the expansion rate of empty space and volume containing mass, is the '**g**' field of the mass. Matter connotes as negative pressure sinkholes. Amalgamation of space and mass takes form as local area-density impedance. Three Dimensional global density functions locally as two dimensional area density. After three centuries, we return to *Newton's Law of Inertia*, to explain his *Law of Gravity*.

Fundamental particles are spatial rotations. The unique combination of energy, size and spatial angular momentum, enables electrons and positrons to exist as stable subatomic units called charge. In their combinatorial guise as muons and tau's, they form the building blocks of nature. In their free form, they exist at the atomic level as the essence of chemical bonding, and in the cosmological arena as the long range electric force.

While the principles here laid down are founded upon functional physics fundamentals, they are not as yet recognized as Standard Theory. In the words of John Wheeler:

*"Perhaps a different theory would reveal
all matter is made of electrons."*